A Guide to Small-Scale Energy Harvesting Techniques

Edited by Reccab Manyala

Published in London, United Kingdom

IntechOpen

Supporting open minds since 2005

A Guide to Small-Scale Energy Harvesting Techniques
http://dx.doi.org/10.5772/intechopen.77520
Edited by Reccab Manyala

Contributors
Sajid Sajid, Haoran Jiang, Jun Ji, Xin Liu, Mingjun Duan, Dong Wei, Peng Cui, Hao Huang, Shangyi Dou, Lihua Chu, Yingfeng Li, Bing Jiang, Meicheng Li, Naser Ahmadi Moghaddam, Alireza Maleki, Akbarov Rasul, Jie Wang, Linglin Zhou, Chunlei Zhang, Zhong Lin Wang, Chris Gould, Rafał Mech, Jerzy Kaleta, Przemysław Wiewiórski

First published in London, United Kingdom, 2020 by IntechOpen
IntechOpen is the global imprint of INTECHOPEN LIMITED, registered in England and Wales, registration number: 11086078, 7th floor, 10 Lower Thames Street, London, EC3R 6AF, United Kingdom
Printed in Croatia

British Library Cataloguing-in-Publication Data
A catalogue record for this book is available from the British Library

Additional hard and PDF copies can be obtained from orders@intechopen.com

A Guide to Small-Scale Energy Harvesting Techniques
Edited by Reccab Manyala
p. cm.
Print ISBN 978-1-78923-909-6
Online ISBN 978-1-78923-910-2
eBook (PDF) ISBN 978-1-83968-506-4

For EU product safety concerns:
IN TECH d.o.o., Prolaz Marije Krucifikse Kozulić 3, 51000 Rijeka, Croatia, info@intechopen.com or visit our website at intechopen.com.

Meet the editor

Reccab Manyala is Professor of Physics at the University of Zambia where he lectures in physics and carries out research and consultancy in new renewable energy and renewable energy technologies. He obtained his PhD in Renewable Energy and Renewable Energy Technologies from Maseno University (Kenya) in 2009. Prof. Manyala holds a Bachelor's degree in Education Science (Physics and Mathematics) and a Master's degree in Theoretical Physics (Plasma Physics and Mathematical Physics) from the University of Nairobi (Kenya), as well as a Master's degree in Experimental Physics (Condensed Matter Physics) from the University of Arkansas (USA). He has published extensively and has research and teaching experience in many areas of physics spanning over 30 years in Kenya and abroad. Prof. Manyala has edited two books for IntechOpen and this is his third.

Contents

Preface

The use of energy can be traced back to humans (species of the genus *Homo*) starting about two million years ago when they started cooking their food using firewood. Cooking had a profound evolutionary effect because it increased food efficiency, which allowed humans to spend less time foraging, chewing, and digesting. Modern anthropologists argue that *Homo erectus* developed a smaller, more efficient digestive tract, which freed up body energy to enable larger brain growth. Extended arguments reveal that cooking and control of fire generally affected species development by providing warmth and helping to fend off predators, which helped human ancestors adapt to a ground-based lifestyle. As humans developed new skills with increased activities, energy interaction and usage emerged. Energy was used not only for domestic functions but also for space applications. With industrialization, humans realized that energy was needed to move machines and do other things as well. In this quest, and without understanding the consequences of using fossil fuels extensively, many problems arose. Researchers in energy embarked on a journey to try to solve some of the problems by studying different forms of renewable energy. To understand different needs, researchers have tried to come up with ways in which small-scale energy harvesting can be adapted to different needs that do not require heavy-duty energy production. Technological advancements point directly to this quest where some gadgets have been miniaturized and others developed to help humans live better lives.

This book attempts to present a number of ideas regarding a few selected small-scale energy harvesting methods and techniques as well as theories and products that may be helpful in improving the quality of life.

Chapter 1 outlines the potential of perovskite solar cells (PSCs) as a promising form of new solar cell for power generation due to their simple processing, abundance of materials, and architectural integration, as well as good power conversion efficiencies, which rocketed from 3.8% in 2009 to 23.3% in 2018. It is pointed out in the chapter that the toxic lead (Pb) element containing the chemical composition of typically used organic–inorganic halide perovskites hinders the practical applications of PSCs. The chapter, however, gives a general discussion on perovskite crystal structure along with serious efforts focused on Pb replacement in these devices. Elaborate fundamental features of tin (Sn)-based perovskites together with their performance in PSCs is then presented, and alternative elements, such as copper (Cu), germanium (Ge), bismuth (Bi), and antimony (Sb), are outlined. Last but not least in the chapter is a summary of the challenges and opportunities based on the chapter.

In Chapter 2, thermoelectric energy generation of electrical power from temperature gradients or differences in naturally occurring geothermal heat and rocks, or from waste heat in man-made equipment and industrial processes, are discussed. Their commercial applications to replace or recharge batteries in low-power electronic systems are presented. The fundamental thermoelectric theory related to power generation, including the theoretical analysis and numerical calculations required to calculate the thermoelectric efficiency and electrical power generated when a single thermoelectric couple is given as an example. The short-term challenge for thermoelectric energy harvesting, which is a cost-effective and practical solution to replace batteries, and can be scaled to provide sufficient power to operate electrical rotating machines such as low-power motors and pumps, is clearly outlined. On the other hand, the long-term challenge of improving the efficiency, power output, cost of thermoelectric modules, and energy harvesting systems, and to develop them from low-power to medium-power applications, is presented.

The work in Chapter 3 focuses on the development of new kinds of energy harvesters that could be used in various applications, including industrial, aerospace, or consumer markets. The main aspect explores transformation of different sources of energy, such as temperature, vibration, shock, etc., into usable electric power, which in normal conditions is wasted. The process of energy harvesting relies on harvesters employing magnetomechanical effects.

Chapter 4 introduces a new energy technology in small-scale energy harvesting in relation to triboelectric nanogenerators (TENGs) that can harvest ambient mechanical energy and convert it to electricity for continuously powering small electronic devices. The fundamental working mechanism and modes of a TENG are presented. The chapter outlines how the technology can harvest all kinds of mechanical energy, especially at low frequencies, such as human motion, walking, vibration, mechanical triggering, rotating tires, wind, moving automobiles, flowing water, rain drops, as well as ocean waves.

Chapter 5 deals with the exploitation of radio waves existing in the ambient environment for battery charging, called radiofrequency energy harvesting. A method based on spectrum sensing to allow wireless devices to select the frequency band with maximum power that exceeds a predefined threshold to charge the device is described. The power threshold can be determined according to battery type and its required charging power, and the device can use this power for battery charging.

Chapter 6 does not seem to fall directly under the subject matter of this book, but covers one of the promising directions in the use of solar energy—heliomaterials science, whether for large-scale or small-scale energy production. It was, therefore, thought necessary to include this chapter in the book. The work describes important characteristics of engineering processes such as capacity, maximum and average energy densities, uniformity of energy density distribution, focal spot size, characteristics of energy density distribution and its change in time, duration of the process, and start and stop mode rate, which must be taken into consideration when developing solar furnaces. The work also looks at the drawbacks of solar concentrators and the variability of the characteristics of the focal spot with time. New ideas on the implementation of such a system are given in the discussion and conclusion.

Reccab Manyala
The University of Zambia,
School of Natural Sciences,
Department of Physics,
Lusaka, Zambia

Chapter 1

Quest for Lead-Free Perovskite-Based Solar Cells

Sajid Sajid, Jun Ji, Haoran Jiang, Xin Liu, Mingjun Duan, Dong Wei, Peng Cui, Hao Huang, Shangyi Dou, Lihua Chu, Yingfeng Li, Bing Jiang and Meicheng Li

Abstract

Today, the perovskite solar cells (PSCs) are showing excellent potentials in terms of simple processing, abundance of materials, and architectural integration, as well as very promising device's power conversion efficiencies (PCEs), rocketed from 3.8% in 2009 to 23.3% in 2018. However, the toxic lead (Pb) element containing the chemical composition of typically used organic-inorganic halide perovskites hinders the practical applications of PSCs. This chapter starts with a general discussion on the perovskite crystal structure along with the serious efforts focused on Pb replacement in these devices. Section 2 will elaborate the fundamental features of tin (Sn)-based perovskites together with their performance in the PSCs. Other alternative elements, such as copper (Cu), germanium (Ge), bismuth (Bi), and antimony (Sb), will be discussed in Section 3. The end will summarize the challenges and opportunities based on the chapter contents.

Keywords: toxicity, stability, lead-free perovskites, chemical composition

1. Introduction

PSCs with organometal (Pb) halide perovskites as photo-absorber showed rapid development in terms of PCEs from 3.8 to 23.3% [1–7]. The typically used Pb-based perovskites possess several appealing advantages such as broadband absorption range, long diffusion length, low exciton binding energy, and high-charge-carrier mobility [2, 3, 8–14]. However, intrinsic toxicity of Pb-based perovskites is a serious issue for both human and environment [15–19]. In this context, the replacement of Pb element in PSCs is extremely important for economical clean energy conversion devices which would benefit mankind in future endeavor.

Organic-inorganic trihalide perovskite is generally represented by ABX_3 ($A = CH_3NH_3^+$ (MA), $CH(NH_2)_2^+$ (FA), Cs^+; $B = Cu^{2+}$, Pb^{2+}, Sn^{2+}, Ge^{2+}, Bi^{3+}, Sb^{3+}; $X = I^-$, Cl^-, Br^-). **Figure 1** illustrates the typical cubic perovskite structure with basic octahedron (BX_6) unit. It has been witnessed that structure distortions determine the physical/electrical properties of ABX_3 perovskite [20]. For example, Goldschmidt's tolerance prediction can be used for the dimensional evaluation of a perovskite as follows:

$$t = \frac{(r_A + r_X)}{\sqrt{2}(r_B + r_X} \quad (1)$$

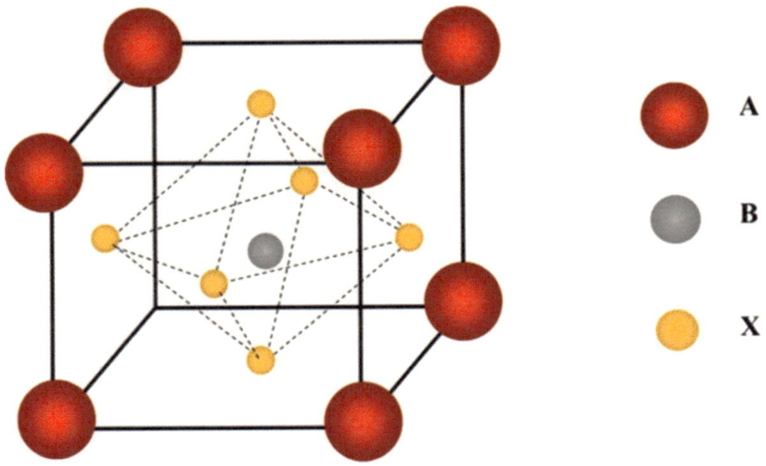

Figure 1.
Basic crystal structure of perovskite semiconductor, where A, B, and X represent ($CH_3NH_3^+$, $CH(NH_2)_2^+$, Cs^+), (Cu^{2+}, Pb^{2+}, Sn^{2+}, Ge^{2+}, Bi^{3+}, Sb^{3+}), and (I^-, Cl^-, Br^-), respectively.

where r_A, r_B, and r_X denote the ionic radii of A, B, and X, respectively. If the value of "t" results in the range between 0.813 and 1.107, then it is considered as a high-symmetry cubic three-dimensional (3D) perovskite, while two-dimensional (2D), one-dimensional (1D), and zero-dimensional (0D) perovskites are formed when "t" gives other values than the abovementioned range [21, 22]. The structural dimensionality approach is considered one of the essential factors because different dimensions of perovskite influence the kinetics of charge carriers. Nonetheless, this evaluation is not enough to be applied to all perovskite semiconductors. Therefore the probe for electronic dimensionality is equally important [23]. For example, perovskite materials with low electronic dimensionality but high structural dimensionality have less promises as light absorbers because of the barrier to isotropic current flow, large effective masses of holes/electrons, and deeper defect states.

The excellent performance of Pb-based perovskites is mainly due to high structural symmetry and strong antibonding coupling between Pb and I [24]. In a similar way, Cu^{2+}, Ge^{2+}, Bi^{3+}, Sb^{3+}, and Sn^{2+} with ns^2 lone pairs could be used with halides to obtain octahedral structure; therefore, they are investigated as alternatives to the toxic Pb element [15, 25, 26]. Herein, we will introduce the Pb-free perovskites from previously reported theoretical calculations and experimental studies.

2. Sn-based perovskites

Tin (Sn) element has been widely used as an alternative to Pb, since both occur in group IVA of the periodic table with similar ionic radii (Pb: 1.49 Å and Sn: 1.35 Å). Therefore, Pb substitution by Sn would cause no obvious lattice distortion in perovskites [27]. The intrinsic instability that results in the decomposition of unstable products such as SnI_2 and HI (acidifier) and toxicologically inactive oxygenated Sn precipitates are still remained the toxicity issues in the Sn-based perovskites [16]. However, benefiting from the stability of PSCs based on Sn perovskites and easy cleaning of Sn from the human body compared to Pb, Sn-based perovskites could be a better choice than perovskite with Pb cation. Density functional theory (DFT) and GW approximation are typically used for the structural and electrical properties of Sn-based perovskites [28, 29]. For instance, the theoretical–experimental calculations about the bandgap of $MASnI_3$ showed a range of 1–1.3 eV, charge mobility of 1.6 $cm^2 V^{-1} s^{-1}$, and diffusion length up to 30 nm (**Figure 2A**) [29, 30, 32, 33]. Additionally, SnF_2-doped $MASnI_3$ showed tenfold larger carrier lifetime with diffusion length of 500 nm (**Figure 2B**) [31]. Replacing $MASnI_3$ by $FASnI_3$ results in a desirable bandgap of 1.41 eV and minimized oxidation of Sn^{2+} to Sn^{4+} [34, 35]. It is also reported that replacement of "A" cation by Cs^+ in $ASnX_3$ yields higher charge mobility, lower exciton binding energy, and large optical absorption coefficient compared to conventional $MAPbI_3$ [36, 37]. However, plasticity of the tin-halide-tin angle and Cs^+ cation migration [38] showed $CsSnX_3$ a phase transitional perovskite with respect to temperature [39].

Since Sn-based perovskites possess low crystallization barrier and high solubility in solvents, their thin films can be fabricated at low temperature in the PSCs. For example, a high crystalline $MASnI_3$-perovskite thin film has been prepared from a transitional SnI_2•3DMSO intermediate phase (**Figure 3A**) [40]. This high-quality perovskite film formation in a hole-selective layer-free PSCs resulted in a photocurrent of 21 mA cm^{-2}. In order to obtain low doping level in $MASnI_3$-perovskite thin film, a low-temperature vapor-assisted solution process was employed (**Figure 3B**), where the excess of Sn^{2+} compounds due to $Sn(OH)_2$ and SnO resulted in low hole-doping level [41, 44]. Furthermore, anti-solvent dripping process was used to control fast crystallization and fabricate pinhole-free thin films of Sn-based perovskites. The diethyl ether dripping on $FASnI_3$ and chlorobenzene on $(FA)_{0.75}(MA)_{0.25}SnI_3$ enabled the as-prepared PSCs to obtain efficiency of 6.2 and 8.1%, respectively (**Figure 3C** and **D**) [42, 43].

It is reported that the oxidation of Sn-based perovskites (self-doping from Sn^{2+} to Sn^{4+}) leads to carrier recombination and poor device performance. In this context, specific amount of SnF_2 can be used as an inhibitor for Sn^{4+}. The large quantity of SnF_2 may generate phase separation such as plate-like aggregates on the perovskite film surface. Here, the strong binding affinity in SnF_2-pyrazine complex was helpful to

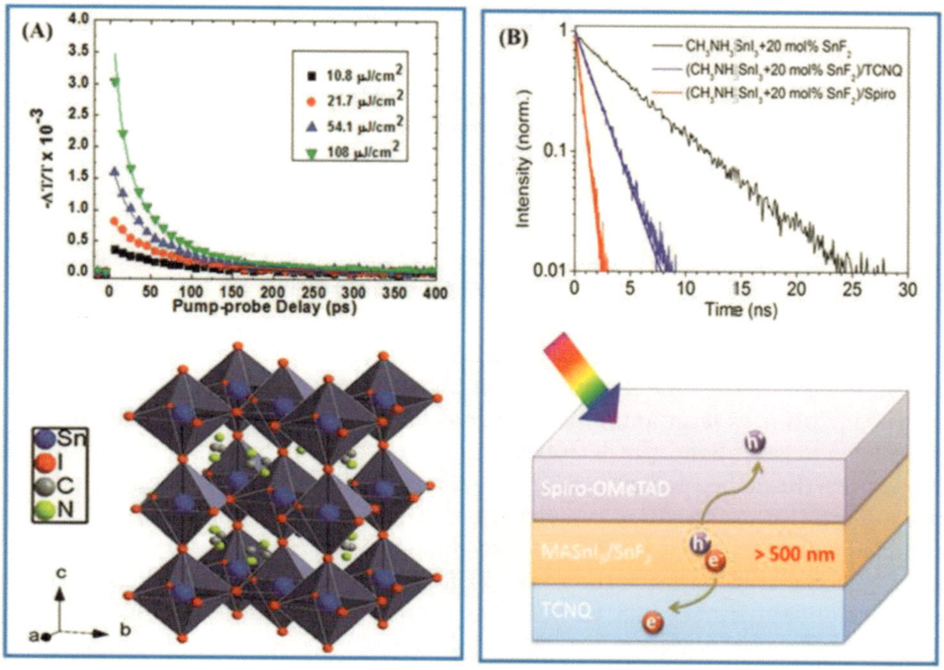

Figure 2.
(A) Transient terahertz photoconductivity and simulated crystal structure of $MASnI_3$. (B) Time-resolved photoluminescence for 20 mol (%) SnF_2-doped $MASnI_3$ thin film and schematic device illustration. Reprinted with permission from [30, 31].

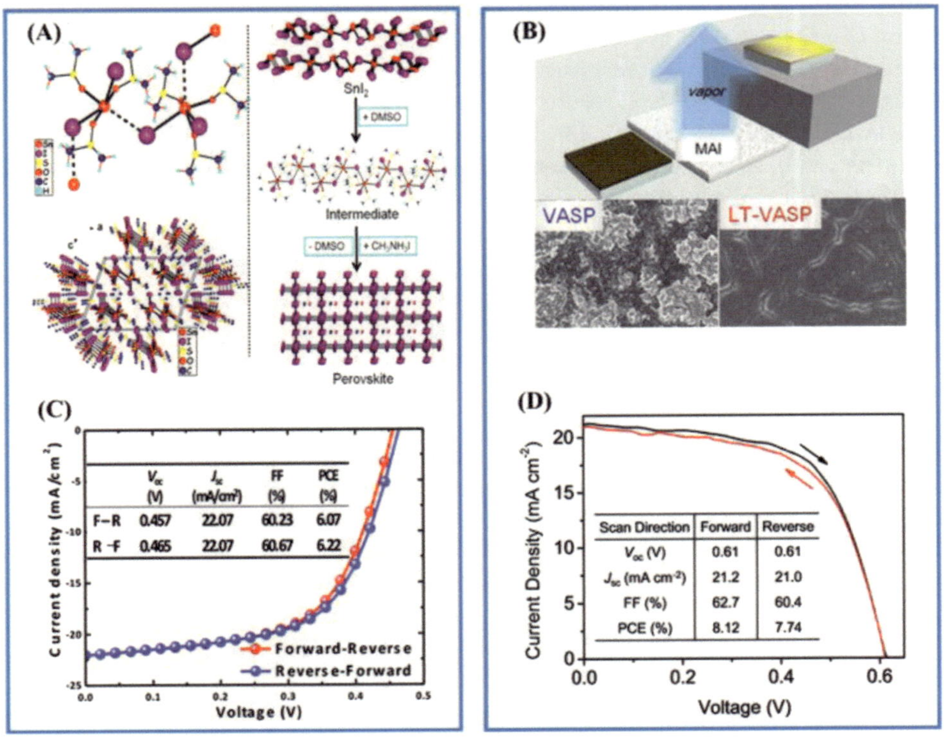

	V_{oc} (V)	J_{sc} (mA/cm^2)	FF (%)	PCE (%)
F-R	0.457	22.07	60.23	6.07
R-F	0.465	22.07	60.67	6.22

Scan Direction	Forward	Reverse
V_{oc} (V)	0.61	0.61
J_{sc} (mA cm^{-2})	21.2	21.0
FF (%)	62.7	60.4
PCE (%)	8.12	7.74

Figure 3.
(A) The schematic illustrations of the $SnI(DMSO)^{3+}$ ions linked with lone I^- ions, unit cell of SnI_2•3DMSO, and the film formation of the $MASnI_3$ perovskite starting from SnI_2 through SnI_2•3DMSO intermediate. (B) Scanning electron microscopy images of $MASnI_3$ thin films obtained through low-temperature vapor-assisted solution process and vapor-assisted solution method. (C) Current–voltage characteristic curves of the cell based on $FASnI_3$ with 10% SnF_2 additives. (D) Current-voltage curves of the best device based on $(FA)_{0.75}(MA)_{0.25}SnI_3$ perovskite. Reprinted with permission from [40–43].

inhibit the phase separation caused by excess SnF_2 (**Figure 4A**) [45]. Furthermore, the built-in potential can also be optimized through SnF_2 and thus high open-circuit voltage of the device by energetic landscape alignments [46] as illustrated in (**Figure 4B**). Besides SnF_2, other additives such as $SnBr_2$, SnI_2, and $SnCl_2$ have also been explored where $SnCl_2$ exhibits the highest stability by inhibiting the decomposition/oxidation [34, 47]. Moreover, the employment of hypophosphorous acid in $CsSnIBr_2$ showed seed-like perovskite where Sn^{2+} oxidation was significantly reduced [48]. As a result, the charge recombination rate was decreased by fourfold compared to the control devices. The as-prepared cells displayed excellent oxygen-moisture stability at ambient conditions and thermal stability in vacuum environment.

The inclusion of large ammonium cations for tuning the dimensionality or generating massive Schottky defects has also been tested in Sn-based perovskites to optimize the device performance. For example, a different ratio of phenylethylammonium (PEA) as a cation can yield perovskites with two-dimensional, three-dimensional, and three-dimensional-two-dimensional-mixed dimensionalities. The high-quality thin film of the mixed PEA-FA perovskite (20% of PEA) delivered a stable device efficiency of 5.9% [49]. The composition of ethylenediammonium (en) and formamidinium (FA) in $FASnI_3$/$MASnI_3$/$CsSnI_3$ crystal indicates no effect on the dimensionality of the perovskite because the size of en is too large for the unit cell cage and it can remove a certain $\{SnI\}^+$ species. Here, the point defects are recognized as the Schottky defects. However, the optimized devices based on {en}$MASnI_3$ and {en}$FASnI_3$ showed efficiency of 6.6 and 7.1%, respectively [50, 51].

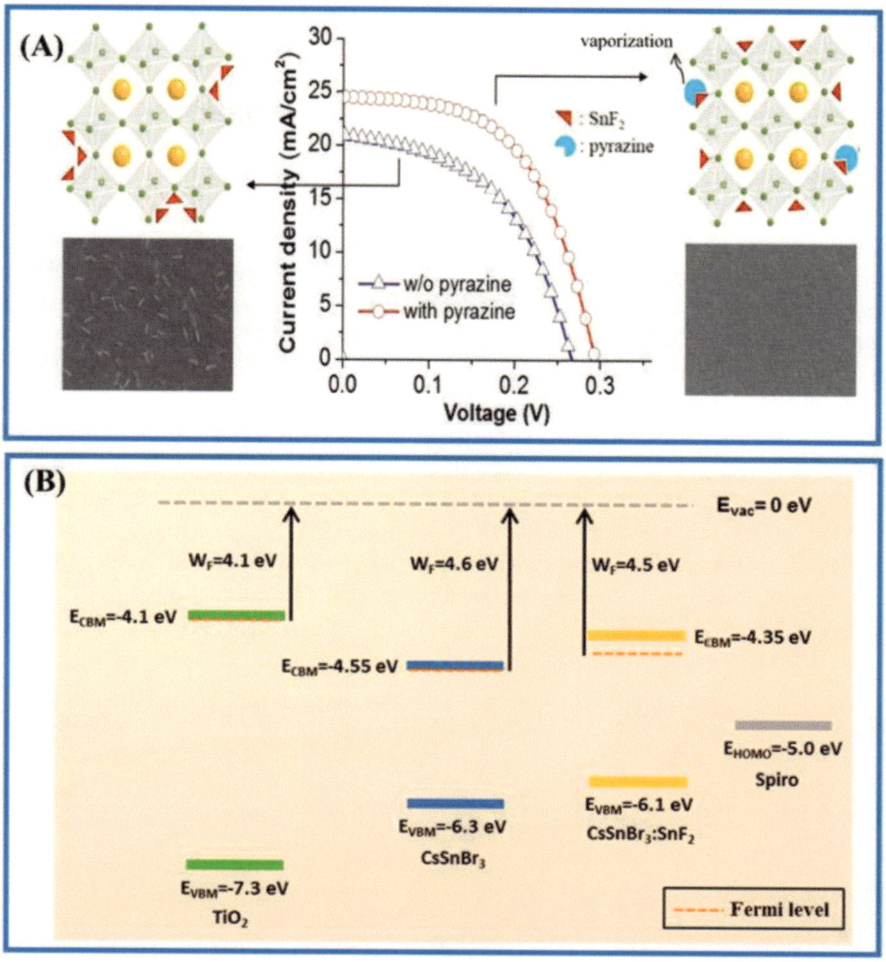

Figure 4.
(A) Schematic crystal structure of $FASnI_3$ with SnF_2 and SnF_2-pyrazine complex with corresponding scanning electron microscopy images. (B) Representation of energetic landscapes of the TiO_2, pure $CsSnBr_3$, and 20 mol (%) of SnF_2 in $CsSnBr_3$ and spiro-OMeTAD with respect to vacuum level. Reprinted with permission from [45, 46].

3. Ge-, Bi-, Sb-, and Cu-based perovskites

Although Sn-based PSC has achieved PCE over 8%, the oxidation of Sn and degradation of the perovskite still need to be addressed. In this context, other metal halide perovskites such as Ge, Bi, Sb, and Cu are considered as potential candidates to replace Pb element in the perovskite crystal. As Ge 4 s has higher orbital energy compared to Pb 6 s and Sn 5 s, thus Ge-based perovskites should exhibit smaller bandgaps. However, in practice the Ge-based perovskites such as $CsGeI_3$, $MAGeI_3$, and $FAGeI_3$ displayed larger bandgaps (i.e., 1.63, 2.0, and 2.35 eV) than that of $CH_3NH_3SnI_3$ (1.30 eV) and $CH_3NH_3PbI_3$ (1.55 eV) [52]. The difference between experimental data and what we expected from high orbital energy is mainly due to the structural distortion of $[GeI_6]$ octahedral as the small ionic radius (0.73 Å) of Ge^{2+} substituting the bigger ionic radius of Pb^{2+} (1.19 Å) or Sn^{2+} (1.02 Å) [53]. Additionally, the Ge-based perovskites crystallize in polar space groups [54]. The Ge^{2+} cation cannot sustain at the center of octahedron and forms three long Ge—I bonds (2.73–2.77 Å) and three short Ge-I bonds (3.26–3.58 Å) [53]. This means that the Ge cation cannot maintain the desirable octahedral crystal structure which may be one of the reasons for poorer device performance (the best PCE up to 0.2% so far) as compared to Pb- or Sn-based perovskite [52].

Furthermore, Bi and Sb can form $A_3B_2X_9$ perovskite structure such as $A_3Bi_2X_9$ and $A_3Sb_2X_9$ [55, 56]. These perovskites typically form low-dimensional perovskites with two polymorphs. Devices fabricated with $Cs_3Bi_2I_9$, $(MA)_3Bi_2I_9$, and $MA_3Bi_2I_{9-x}Cl_x$ (hexagonal phase and *P*63/mm(194) space group) delivered PCE of 1, 0.33, and 0.38%, respectively [35]. The highest performance in the case of $Cs_3Bi_2I_9$ perovskite was attributed to the low non-radiative recombination. The A cation has also a crucial role in the formation of $Cs_3Sb_2I_9$ perovskite. For instance, the mixture of Cs^+ and MA^+ cations forms dimer phases [57], which is structurally different phase from rubidium (Rb) as the A cation. The DFT analysis about the dimer and layer forms of $A_3Sb_2I_9$ (A = Cs or Rb) indicated much preference of Rb-based systems (with formation energy of 0.25 eV) as layered phase compared to Cs-based perovskite (formation energy of 0.1 eV). Thus the $Rb_3Sb_2I_9$ perovskite in PSC showed a PCE of 0.66% [57]. Recently, our research group fabricated Cu-based hybrid materials denoted as $(MA)_2CuX$ (where X = Cl_4, I_2Cl_2, or Br_2Cl_2) to replace Pb [15]. It was revealed that chlorine (Cl^-) in the perovskite crystal has a critical role in the stabilization of the as-prepared materials. The corresponding PCEs of the devices are depicted in **Figure 5**.

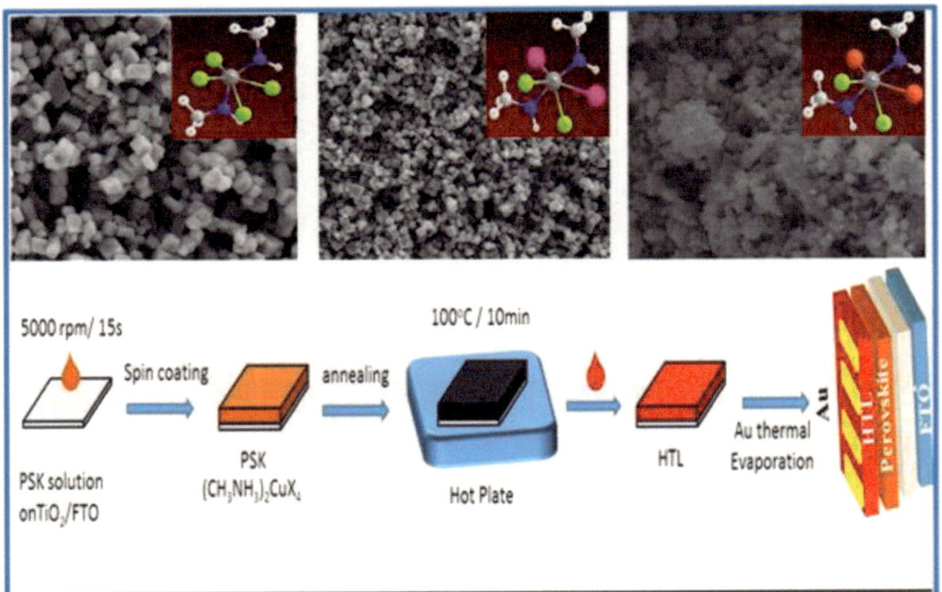

Samples	J_{SC} /mA cm^{-2}	V_{OC} /V	FF	η /%
$(CH_3NH_3)_2CuCl_4$	8.12	0.560	52.99	2.41
$(CH_3NH_3)_2CuCl_2I_2$	6.78	0.545	47.41	1.75
$(CH_3NH_3)_2CuCl_2Br_2$	3.35	0.581	50.86	0.99

Figure 5.
SEM micrographs of $(MA)_2CuCl_4$, $(MA)_2CuI_2Cl_2$, and $(MA)_2CuBr_2Cl_2$ with their chemical structures, respectively. Schematic illustration of the as-fabricated devices along with photovoltaic parameter table. Reprinted with permission from [35].

4. Summary and future outlook

It is concluded that the nontoxic elements should form octahedral crystal structure with halides. Besides the bandgaps, structural/electronic dimensionality, crystal defects, carrier motilities, and high-quality film formation of Pb-free perovskites are equally important. Although Sn is widely used to replace Pb in perovskites, it oxidizes from Sn^{2+} to Sn^{4+} and leads to high charge recombination and poor device performance. In addition, the fast crystallization of Sn-based perovskites forms pores in their thin films that deteriorate the solar cell performance. Thus, additives and additional solvent such as SnF_2, $SnCl_2$, SnF_2-pyrazine complex, $SnI_2(DMSO)_x$ complex, and HPA can be optimized to inhibit the oxidation reaction and control film growth rate. Another severe effect found in the Sn-based device is the hysteresis that is mostly associated with the imbalance charge transport and defects. In this context, mitigation of ion migration within perovskite and interfacial engineering by using suitable charge-transporting materials can reduce the hysteresis effect.

The Ge-based perovskites exhibit larger bandgaps which can deliver high open-circuit voltage compared to Sn-based perovskite. However, they also show severe oxidation. Alternatively, perovskite structure with trivalent Bi^{3+}/Sb^{3+} cations can be designed. Copper is also a promising alternative to Pb in mixed halide perovskites where the Cu reduction can be decreased and the material stability as well as the perovskite crystallization can be enhanced by manipulating the halide ions. Replacing the toxic Pb in perovskite crystals with environmentally friendly, nontoxic, earth-abundant, and cost-effective materials such as transition metals (Fe^{2+}, Zn^{2+}, Cu^{2+}, etc.) is an important target for sustainable energy perspectives.

Acknowledgements

This work is supported partially by the National Natural Science Foundation of China (Grant nos. 51772096), Natural Science Foundation of Beijing Municipality (L172036), Joint Funds of the Equipment Pre-Research and Ministry of Education (6141A020225), Par-Eu Scholars Program, Science and Technology Beijing 100 Leading Talent Training Project, Beijing Municipal Science and Technology Project (Z161100002616039), Fundamental Research Funds for the Central Universities (2016JQ01, 2017ZZD02), and NCEPU "Double First-Class" Graduate Talent Cultivation Program.

Conflict of interest

The authors declare no conflict of interest.

Author details

Sajid Sajid, Jun Ji, Haoran Jiang, Xin Liu, Mingjun Duan, Dong Wei, Peng Cui, Hao Huang, Shangyi Dou, Lihua Chu, Yingfeng Li, Bing Jiang and Meicheng Li*
State Key Laboratory of Alternate Electrical Power System with Renewable Energy Sources, School of Renewable Energy, North China Electric Power University, Beijing, China

*Address all correspondence to: mcli@ncepu.edu.cn

References

[1] Kojima A, Teshima K, Shirai Y, Miyasaka T. Organometal halide perovskites as visible-light sensitizers for photovoltaic cells. Journal of the American Chemical Society. 2009;**131**(17):6050-6051

[2] Cui P, Wei D, Ji J, et al. Highly efficient electron-selective layer free perovskite solar cells by constructing effective p–n heterojunction. Solar RRL. 2017;**1**(2):1600027

[3] Wei D, Ji J, Song D, et al. A TiO_2 embedded structure for perovskite solar cells with anomalous grain growth and effective electron extraction. Journal of Materials Chemistry A. 2017;**5**(4):1406-1414

[4] Wei D, Ma F, Wang R, et al. Ion-migration inhibition by the cation–π interaction in perovskite materials for efficient and stable perovskite solar cells. Advanced Materials. 2018;**30**(31):1707583

[5] Sajid S, Elseman AM, Huang H, et al. Breakthroughs in NiO_x-HTMs towards stable, low-cost and efficient perovskite solar cells. Nano Energy. 2018;**51**:408-424

[6] Elseman AM, Sharmoukh W, Sajid S, et al. Superior stability and efficiency over 20% perovskite solar cells achieved by a novel molecularly engineered Rutin–AgNPs/thiophene copolymer. Advanced Science. 2018;**5**(11):1800568

[7] Available from: https://www.nrel.gov/pv/assets/pdfs/pv-efficiencies-07-17-2018.pdf

[8] Song D, Cui P, Wang T, et al. Managing carrier lifetime and doping property of lead halide perovskite by postannealing processes for highly efficient perovskite solar cells. The Journal of Physical Chemistry C. 2015;**119**(40):22812-22819

[9] Elseman A, Ji J, Dou S, et al. Novel hole transport layer of nickel oxide composite with carbon for high-performance perovskite solar cells. Chinese Physics B. 2018;**27**(1):017305

[10] Sajid S, Elseman AM, Ji J, et al. Computational study of ternary devices: Stable, low-cost, and efficient planar perovskite solar cells. Nano-Micro Letters. 2018;**10**(3):51

[11] Elseman AM, Sajid DW, Shalan AE, Rashad MM, Li M. Pathways towards high-stable, low-cost and efficient perovskite solar cells. In: Emerging Solar Energy Materials. IntechOpen; 2018

[12] Yue S, Liu K, Xu R, et al. Efficacious engineering on charge extraction for realizing highly efficient perovskite solar cells. Energy & Environmental Science. 2017;**10**(12):2570-2578

[13] Zhang Z, Yue X, Wei D, et al. DMSO-based PbI_2 precursor with $PbCl_2$ additive for highly efficient perovskite solar cells fabricated at low temperature. RSC Advances. 2015;**5**(127):104606-104611

[14] Elseman AM, Rashad MM, Hassan AM. Easily attainable, efficient solar cell with mass yield of nanorod single-crystalline organo-metal halide perovskite based on a ball milling technique. ACS Sustainable Chemistry & Engineering. 2016;**4**(9):4875-4886

[15] Elseman AM, Shalan AE, Sajid S, Rashad MM, Hassan AM, Li M. Copper-substituted lead perovskite materials constructed with different halides for working $(CH_3NH_3)_2CuX_4$-based perovskite solar cells from experimental and theoretical view. ACS Applied Materials & Interfaces. 2018;**10**(14):11699-11707

[16] Babayigit A, Ethirajan A, Muller M, Conings B. Toxicity of organometal

halide perovskite solar cells. Nature Materials. 2016;**15**:247

[17] Kadro JM, Pellet N, Giordano F, et al. Proof-of-concept for facile perovskite solar cell recycling. Energy & Environmental Science. 2016;**9**(10):3172-3179

[18] Sajid EAM, Ji J, et al. Novel hole transport layer of nickel oxide composite with carbon for high-performance perovskite solar cells. Chinese Physics B. 2018;**27**(1):17305-017305

[19] Sajid S, Elseman AM, Wei D, et al. NiO@Carbon spheres: A promising composite electrode for scalable fabrication of planar perovskite solar cells at low cost. Nano Energy. 2018/11/03/

[20] Gao P, Grätzel M, Nazeeruddin MK. Organohalide lead perovskites for photovoltaic applications. Energy & Environmental Science. 2014;**7**(8):2448-2463

[21] Zhao Y, Zhu K. Organic-inorganic hybrid lead halide perovskites for optoelectronic and electronic applications. Chemical Society Reviews. 2016;**45**(3):655-689

[22] Tsai H, Nie W, Blancon J-C, et al. High-efficiency two-dimensional Ruddlesden–Popper perovskite solar cells. Nature. 2016;**536**:312

[23] Xiao Z, Meng W, Wang J, Mitzi DB, Yan Y. Searching for promising new perovskite-based photovoltaic absorbers: The importance of electronic dimensionality. Materials Horizons. 2017;**4**(2):206-216

[24] Yin W-J, Shi T, Yan Y. Unique properties of halide perovskites as possible origins of the superior solar cell performance. Advanced Materials. 2014;**26**(27):4653-4658

[25] Ganose AM, Savory CN, Scanlon DO. Beyond methylammonium lead iodide: Prospects for the emergent field of ns(2) containing solar absorbers. Chemical Communications. 2016;**53**(1):20-44

[26] Konstantakou M, Stergiopoulos T. A critical review on tin halide perovskite solar cells. Journal of Materials Chemistry A. 2017;**5**(23):11518-11549

[27] Zuo F, Williams ST, Liang P-W, Chueh C-C, Liao C-Y, Jen AK-Y. Binary-metal perovskites toward high-performance planar-heterojunction hybrid solar cells. Advanced Materials. 2014;**26**(37):6454-6460

[28] Bernal C, Yang K. First-principles hybrid functional study of the organic–inorganic perovskites $CH_3NH_3SnBr_3$ and $CH_3NH_3SnI_3$. The Journal of Physical Chemistry C. 2014;**118**(42):24383-24388

[29] Umari P, Mosconi E, De Angelis F. Relativistic GW calculations on $CH_3NH_3PbI_3$ and $CH_3NH_3SnI_3$ perovskites for solar cell applications. Scientific Reports. 2014;**4**:4467

[30] Noel NK, Stranks SD, Abate A, et al. Lead-free organic-inorganic tin halide perovskites for photovoltaic applications. Energy & Environmental Science. 2014;**7**(9):3061-3068

[31] Ma L, Hao F, Stoumpos CC, Phelan BT, Wasielewski MR, Kanatzidis MG. Carrier diffusion lengths of over 500 nm in lead-free perovskite $CH_3NH_3SnI_3$ films. Journal of the American Chemical Society. 2016;**138**(44):14750-14755

[32] Chiarella F, Zappettini A, Licci F, et al. Combined experimental and theoretical investigation of optical, structural, and electronic properties of $CH_3NH_3SnX_3$ thin films(X = Cl, Br). Physical Review B. 2008;**77**(4):045129

[33] Fujihara T, Terakawa S, Matsushima T, Qin C, Yahiro M,

Adachi C. Fabrication of high coverage $MASnI_3$ perovskite films for stable, planar heterojunction solar cells. Journal of Materials Chemistry C. 2017;**5**(5):1121-1127

[34] Tsai C-M, Mohanta N, Wang C-Y, et al. Formation of stable tin perovskites co-crystallized with three halides for carbon-based mesoscopic lead-free perovskite solar cells. Angewandte Chemie. 2017;**129**(44):14007-14011

[35] Park B-W, Philippe B, Zhang X, Rensmo H, Boschloo G, Johansson EMJ. Bismuth based hybrid perovskites $A_3Bi_2I_9$ (A: methylammonium or cesium) for solar cell application. Advanced Materials. 2015;**27**(43):6806-6813

[36] Chen Z, Yu C, Shum K, et al. Photoluminescence study of polycrystalline $CsSnI_3$ thin films: Determination of exciton binding energy. Journal of Luminescence. 2012;**132**(2):345-349

[37] Shum K, Chen Z, Qureshi J, et al. Synthesis and characterization of $CsSnI_3$ thin films. Applied Physics Letters. 2010;**96**(22):221903

[38] Chung I, Song J-H, Im J, et al. $CsSnI_3$: Semiconductor or metal? High electrical conductivity and strong near-infrared photoluminescence from a single material. High hole mobility and phase-transitions. Journal of the American Chemical Society. 2012;**134**(20):8579-8587

[39] Liu C, Li W, Fan J, Mai Y. A brief review on the lead element substitution in perovskite solar cells. Journal of Energy Chemistry. 2018;**27**(4):1054-1066

[40] Hao F, Stoumpos CC, Guo P, et al. Solvent-mediated crystallization of $CH_3NH_3SnI_3$ films for heterojunction depleted perovskite solar cells. Journal of the American Chemical Society. 2015;**137**(35):11445-11452

[41] Yokoyama T, Cao DH, Stoumpos CC, et al. Overcoming short-circuit in lead-free $CH_3NH_3SnI_3$ perovskite solar cells via kinetically controlled gas–solid reaction film fabrication process. The Journal of Physical Chemistry Letters. 2016;**7**(5):776-782

[42] Liao W, Zhao D, Yu Y, et al. Lead-free inverted planar formamidinium tin triiodide perovskite solar cells achieving power conversion efficiencies up to 6.22%. Advanced Materials. 2016;**28**(42):9333-9340

[43] Zhao Z, Gu F, Li Y, et al. Mixed-organic-cation tin iodide for lead-free perovskite solar cells with an efficiency of 8.12%. Advanced Science. 2017;**4**(11):1700204

[44] Yokoyama T, Song T-B, Cao DH, Stoumpos CC, Aramaki S, Kanatzidis MG. The origin of lower hole carrier concentration in methylammonium tin halide films grown by a vapor-assisted solution process. ACS Energy Letters. 2017;**2**(1):22-28

[45] Lee SJ, Shin SS, Kim YC, et al. Fabrication of efficient formamidinium tin iodide perovskite solar cells through SnF_2-pyrazine complex. Journal of the American Chemical Society. 2016;**138**(12):3974-3977

[46] Gupta S, Bendikov T, Hodes G, Cahen D. $CsSnBr_3$, a lead-free halide perovskite for long-term solar cell application: Insights on SnF_2 addition. ACS Energy Letters. 2016;**1**(5):1028-1033

[47] Marshall KP, Walker M, Walton RI, Hatton RA. Enhanced stability and efficiency in hole-transport-layer-free $CsSnI_3$ perovskite photovoltaics. Nature Energy. 2016;**1**:16178

[48] Li W, Li J, Li J, Fan J, Mai Y, Wang L. Addictive-assisted construction of all-inorganic $CsSnIBr_2$ mesoscopic perovskite solar cells with superior

thermal stability up to 473 K. Journal of Materials Chemistry A. 2016;**4**(43):17104-17110

[49] Liao Y, Liu H, Zhou W, et al. Highly oriented low-dimensional tin halide perovskites with enhanced stability and photovoltaic performance. Journal of the American Chemical Society. 2017;**139**(19):6693-6699

[50] Ke W, Stoumpos CC, Zhu M, et al. Enhanced photovoltaic performance and stability with a new type of hollow 3D perovskite {en}$FASnI_3$. Science Advances. 2017;**3**(8):1701293

[51] Ke W, Stoumpos CC, Spanopoulos I, et al. Efficient lead-free solar cells based on hollow {en}$MASnI_3$ perovskites. Journal of the American Chemical Society. 2017;**139**(41):14800-14806

[52] Krishnamoorthy T, Ding H, Yan C, et al. Lead-free germanium iodide perovskite materials for photovoltaic applications. Journal of Materials Chemistry A. 2015;**3**(47):23829-23832

[53] Lu X, Zhao Z, Li K, et al. First-principles insight into the photoelectronic properties of Ge-based perovskites. RSC Advances. 2016;**6**(90):86976-86981

[54] Lin Z-G, Tang L-C, Chou C-P. Characterization and properties of novel infrared nonlinear optical crystal $CsGe(Br_xCl_{1-x})_3$. Inorganic Chemistry. 2008;**47**(7):2362-2367

[55] Zhang X, Wu G, Gu Z, et al. Active-layer evolution and efficiency improvement of $(CH_3NH_3)_3Bi_2I_9$-based solar cell on TiO_2-deposited ITO substrate. Nano Research. 2016;**9**(10):2921-2930

[56] Saparov B, Hong F, Sun J-P, et al. Thin-film preparation and characterization of $Cs_3Sb_2I_9$: A lead-free layered perovskite semiconductor. Chemistry of Materials. 2015;**27**(16):5622-5632

[57] Harikesh PC, Mulmudi HK, Ghosh B, et al. Rb as an alternative cation for templating inorganic lead-free perovskites for solution processed photovoltaics. Chemistry of Materials. 2016;**28**(20):7496-7504

Chapter 2

Thermoelectric Energy Harvesting

Chris Gould

Abstract

Thermoelectricity can be used to generate electrical power from temperature gradients or differences in naturally occurring geothermal heat and rocks, or from waste heat in man-made equipment and industrial processes. Thermoelectric energy harvesting systems are finding commercial applications to replace or recharge batteries in low power electronic systems. This chapter provides the fundamental thermoelectric theory related to power generation, including the theoretical analysis and numerical calculations required to calculate the thermoelectric efficiency and electrical power generated when a single thermoelectric couple, and a 127 couple thermoelectric module, are subject to different temperature gradients. A thermoelectric energy harvesting system, incorporating a low power boost converter and DC to DC converter, coupled with electrical energy storage in supercapacitors, is presented and enables a thermoelectric energy harvesting system to provide sufficient electrical power to operate low power electronic components and systems. The short-term challenge for thermoelectric energy harvesting is to become a cost effective and practical solution to replace batteries, and to be scaled to provide sufficient power to operate electrical rotating machines such as low power motors and pumps. The long-term challenge is to improve the efficiency, power output, and cost of thermoelectric modules and energy harvesting systems, and to develop from low power to low-to-medium power applications.

Keywords: thermoelectric, Seebeck, temperature difference, temperature gradient, thermal power generation

1. Introduction

Energy harvesting is an ideal platform to foster research and the commercial application of thermoelectric power generation. The use of naturally occurring temperature gradients or differences found in geothermal heat and rocks, or by man-made waste heat in machinery and industrial processes, can be used to generate electrical power by thermoelectricity. The concept of using thermoelectricity to generate electrical power has been discussed for some time, and is considered to be an environmentally friendly and renewable technology, although thermoelectricity is often overlooked in discussions surrounding renewable energy sources, partly due to the relatively low levels of electrical power generated from a thermoelectric module, which is typically in the milliwatt to watt range, and the low conversion efficiency of between 5 and 10% [1]. However, with the addition of power electronics, coupled with electrical energy storage in electric double layer capacitors, also known as supercapacitors, the instantaneous electrical power output from a thermoelectric power generation system can be increased to a useful level, and can output sufficient electrical power to operate low power electronic systems, recharge or replace batteries in many applications. Furthermore, thermoelectricity can be used in applications where other energy harvesting techniques could not be used, i.e., where light is not available for photovoltaic power generation, or wind for electromagnetic generation, or can be used in combination with other energy harvesting technologies in order to enhance a systems overall performance. The technology is not limited to low power applications, with an on-going focus and research into thermoelectric power generation from waste heat in the automotive market, and is extensively used to provide power to deep-space spacecraft.

The technology has several advantages when used for power generation; thermoelectric modules can function in harsh environments; are relatively small in size and weight; there are no moving parts and very low, if any, maintenance requirements; electrically quiet in operation; do not import dust or other particles; can be oriented in any direction; and the same module can be used for power generation, cooling and heating. The main disadvantage of thermoelectricity is the relatively low conversion efficiency and thermoelectric figure of merit Z_T.

Thermoelectricity has undergone stages of significant interest, research and development, along with periods of inactivity and decline. The scientific principle and potential application of thermoelectric power generation has been known for some time and can be described as the generation of electrical power, via the Seebeck effect, when two dissimilar conducting materials are connected together at one end and subject to a temperature gradient or temperature difference. The fundamental scientific discoveries applicable to thermoelectricity were discovered in the 1800s, with the most important for power generation being the Seebeck effect discovered by Thomas Seebeck in 1821. It should be noted that thermoelectricity can also be used for cooling and heating applications, where a source of DC power is applied to a thermoelectric couple or module's input terminals, resulting in one side of the couple or module reducing in temperature and the other side increasing in temperature and acting as a heat pump. This cooling and heating effect is primarily due to the Peltier effect, discovered in 1834 by Joseph Peltier, and to a lesser extent the Thomson effect in 1855 by William Thomson, later known as Lord Kelvin. Recognition should also be made to Alessandro Volta as an early pioneer in thermoelectric research. The technology developed slowly until the 1930s, when rapid improvements in all areas of thermoelectricity occurred and by the mid 1960s, practical thermoelectric devices emerged for niche applications in aerospace cooling and spacecraft power. Progress in efficiency improvement slowed and research peaked in 1963, followed by a steep decline in activity that was to continue for nearly three decades [2]. However, around 1990 there was renewed interest in thermoelectricity due to a combination of factors, notably environmental concerns regarding refrigerant fluids, alternative refrigeration and interest in cooling electronics [3]. Contemporary interest in the technology is driven by an increasing awareness of the effect of climate change on the planet's environment, a renewed requirement for long-life electrical power sources and energy harvesting technologies, and the increasing miniaturization of electronic circuits and sensors [4]. In recent years, interest has grown in the use of ambient energy sources to power low power electronic systems, with thermoelectricity being one of the most promising and applicable energy harvesting technologies for commercial exploitation.

2. Background thermoelectric theory

This section will present the fundamental thermoelectric theory related to thermoelectric power generation for a single thermoelectric couple, and a 127 couple thermoelectric module.

DOI: http://dx.doi.org/10.5772/intechopen.85670

2.1 Standard thermoelectric couple and module construction

A single thermoelectric couple is constructed from two 'pellets' of semiconductor material usually made from bismuth telluride (Bi_2Te_3), as this material has been found to show the most pronounced thermoelectric effects around room temperature. One of these pellets is doped with acceptor impurity to create a p-type pellet, the other is doped with donor impurity to produce an n-type pellet. The two pellets are physically linked together on one side, usually with a small strip of copper, and mounted between two ceramic outer plates that provide electrical isolation and structural integrity. A single thermoelectric couple, as shown in **Figure 1**, is generally of limited practical use, as the rate of useful power generated due to the Seebeck effect is very small. Practical thermoelectric modules are constructed with several of these thermoelectric couples connected electrically in series and thermally in parallel, with modules typically containing a minimum of three thermoelectric couples, as shown in **Figure 2**, rising to 127 couples for larger devices [5].

2.2 The principle of thermoelectric power generation for a single thermoelectric couple

If a temperature difference is maintained between two sides of a thermoelectric couple, thermal energy will move through the p-type and n-type pellets. As these pellets are electrically conductive, charge carries are transported by this heat. This movement of heat and charge carriers creates an electrical voltage called the Seebeck voltage. If a resistive load is connected across the thermoelectric couple's output terminals, current will flow in the load and an electrical voltage will be generated at the load [6]. A schematic diagram of a single thermoelectric couple, configured for thermoelectric power generation, with the output terminals of the couple connected to a volt meter in order to measure the open-circuit voltage V_{oc} is shown in **Figure 3(a)**, and connected to a resistive load R_L in **Figure 3(b)**.

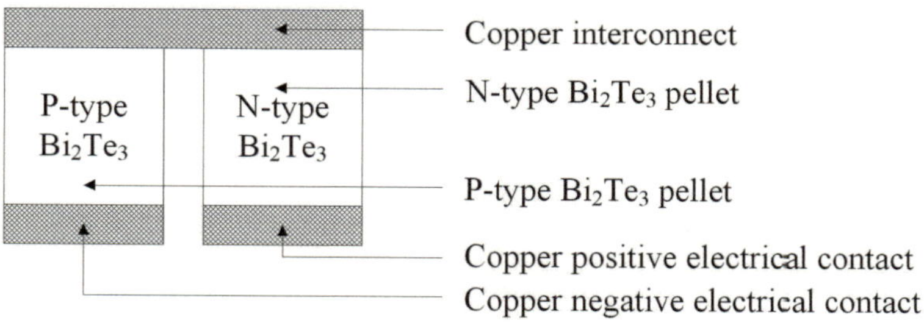

Figure 1.
A single thermoelectric couple.

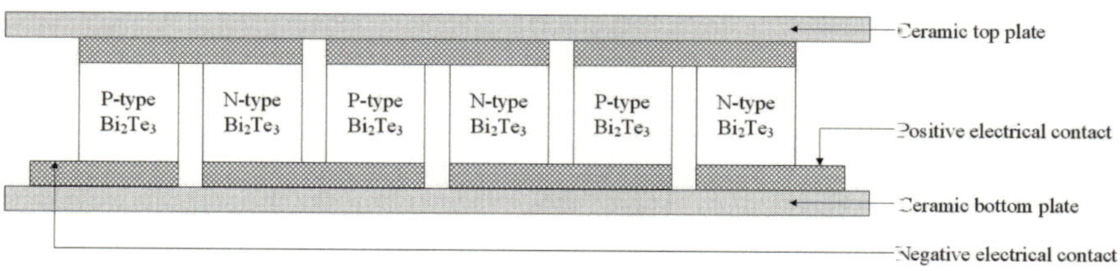

Figure 2.
A three couple thermoelectric module.

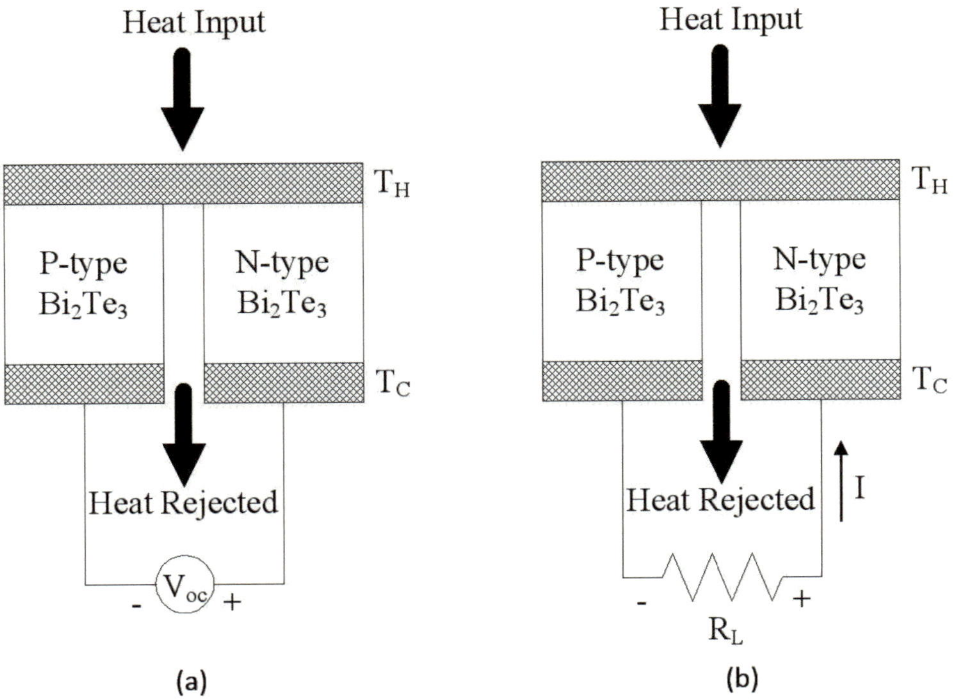

Figure 3.
A single thermoelectric couple configured as a thermoelectric generator with a volt meter measuring the open-circuit voltage V_{oc} (a) and in (b) with a load resistor R_L connected across the couple's output terminals [7].

2.2.1 Calculating the thermoelectric couple open-circuit voltage V_{oc}

According to [8], if we consider the thermoelectric couple in **Figure 3** (a), and assume that no other heat arrives at the cold side T_c other than through the two thermoelectric legs, the thermoelectric couple open-circuit voltage V_{oc} can be found by:

$$V_{oc} = (\alpha_p - \alpha_n)(T_h - T_c) \text{ measured in volts} \tag{1}$$

where α_p and α_n is the Seebeck coefficient of the p-type and n-type pellets of the thermoelectric couple respectively, and may be found quoted in manufacturer literature or obtained by practical measurement. Typical values for Bi_2Te_3 thermoelectric couples are α_p of around 230×10^{-6} V/K and α_n of around -195×10^{-6} V/K [9]. T_h is the surface temperature of the hot side of the thermoelectric couple, and T_c is the surface temperature of the cold side of the thermoelectric couple, measured in Kelvin.

2.2.2 Calculating the resistance R_p and R_n of the thermoelectric couple pellets

The electrical resistivity ρ of each pellet can be found and/or measured, with a typical resistivity for p-type Bi_2Te_3 thermoelectric pellets of around 1.75×10^{-3} Ohms-centimetre, and for n-type Bi_2Te_3 pellets of around 1.35×10^{-3} Ohms-centimetre [10]. The resistance R_p of the p-type pellet can then be found by:

$$R_p = \frac{L_p \rho_p}{A_p} \text{ measured in Ohms} \tag{2}$$

where ρ_p is the electrical resistivity of the pellet in Ohms-cm, L_p is the length of the pellet in cm, and A_p is the cross-sectional area of the pellet in cm^2 found by:

$$A_p = width \times depth \text{ measured in cm}^2 \tag{3}$$

Similarly, the resistance of the n-type pellet R_n can be found by:

$$R_n = \frac{L_n \rho_n}{A_n} \text{ measured in Ohms} \tag{4}$$

2.2.3 The thermoelectric couple resistance R_c

The thermoelectric couple resistance R_c, ignoring the resistance of the copper interconnects, is the addition of the resistance of the two thermoelectric pellets:

$$R_c = R_p + R_n \text{ measured in Ohms} \tag{5}$$

The resistance of the copper interconnects is typically in the micro ohm region, and compared with the resistance of the Bi_2Te_3 thermoelectric pellets, which tends to be in the milliohm region, is significantly smaller in value and hence can be considered as negligible.

2.2.4 Calculating the thermal conductivity K_p and K_n of the thermoelectric couple thermoelements (legs)

The thermal conductivity λ of each pellet can be found and/or measured, with a typical conductivity for a p-type Bi_2Te_3 thermoelectric pellet of around 1.2×10^{-3} W/cm K, and for an n-type Bi_2Te_3 pellet of 1.4×10^{-3} W/cm K [10]. The thermal conductivity K_p of the p-type thermoelement can be found by:

$$K_p = \frac{\lambda_p A_p}{L_p} \text{ measured in watts cm kelvin} \tag{6}$$

where λ_p is the thermal conductivity of the pellet in watts/cm kelvin, L_p is the length of the pellet in cm, and A_p is the cross-sectional area of the pellet in cm^2. Similarly, the thermal conductivity of the n-type pellet K_n can be found by:

$$K_n = \frac{\lambda_n A_n}{L_n} \text{ measured in watts cm kelvin} \tag{7}$$

It should be noted that the electrical resistivity ρ is the reciprocal of the electrical conductivity σ and can also be found using:

$$\sigma = \frac{I \times L}{V \times A} \text{ measured in siemens} \tag{8}$$

where I is the current through a constant cross-sectional area A and length L when a voltage V is applied [8].

2.2.5 Calculating the current I when a load resistor R_L is connected across the thermoelectric couple output terminals

If a load resistance R_L is now connected across the output terminals of the thermoelectric couple, as shown in **Figure 3(b)**, the current I can be expressed as:

$$I = \frac{V}{R_T} \text{ measured in amperes} \tag{9}$$

where V is the closed-circuit voltage, and R_T is the total resistance of the thermoelectric couple and the load. It is possible to find R_T using:

$$R_T = R_p + R_n + R_L \text{ measured in Ohms} \tag{10}$$

Using (Eq. (1)) and (Eq. (10)), the current I can be found by:

$$I = \frac{(\alpha_p - \alpha_n)(T_h - T_c)}{R_p + R_n + R_L} \text{ measured in amperes} \tag{11}$$

2.2.6 The load current I_L and the voltage generated at the load V_L

As the load resistor R_L is connected across the output terminals of the couple, the load current I_L is the same as the circuit current I, and the voltage generated at the load V_L can be found by:

$$V_L = I_L \times R_L \text{ measured in volts} \tag{12}$$

2.2.7 The electrical power generated at the load P_L

The electrical power generated at the load P_L can be found using:

$$P_L = I^2 R_L \text{ measured in watts} \tag{13}$$

If we substitute (Eq. (11)) into (Eq. (13)), we obtain:

$$P_L = \left\{\frac{(\alpha_p - \alpha_n)(T_h - T_c)}{R_p + R_n + R_L}\right\}^2 R_L \text{ measured in watts} \tag{14}$$

Alternatively, the electrical power generated at the load can be found by:

$$P_L = V_L \times I_L \text{ measured in watts} \tag{15}$$

In electrical and electronic engineering, the maximum power transfer between the generator and the load normally occurs when the load resistance is equal to the generator resistance, and if we consider the thermoelectric couple as an ideal model, with no heat lost through thermal radiation or conduction, and ignoring any effect of contact resistances, the maximum efficiency of a thermoelectric generator will not exceed 50%. It should be noted that if the load resistance is increased away from the value that gives rise to maximum power transfer, the power output of the thermoelectric generator will be reduced [8].

2.2.8 Calculating the thermoelectric couple efficiency

The efficiency η of a thermoelectric couple can be found by:

$$\eta = \frac{\textit{Energy supplied to the load}}{\textit{Heat energy absorbed at the hot junction}} \tag{16}$$

The electrical power supplied to the load is P_L, and we now need to find the heat energy absorbed at the hot junction. A proportion of the heat that is absorbed at the hot junction from the heat source T_h is used to balance the Peltier cooling effect in the thermoelectric couple, and an opposing flow of heat occurs due to the thermal conduction of the thermoelectric legs, and by Joule heating within the device. The cooling effect at the source Q_{source} can be found by:

$$Q_{source} = (\alpha_p - \alpha_n) I_L T_h \text{ measured in watts} \tag{17}$$

The opposing heat by conduction in the thermoelectric legs (or pellets) can be found by:

$$Q_{legs} = (T_h - T_c)(K_p + K_n) \text{ measured in watts} \tag{18}$$

where K_p and K_n are the thermal conductance of the p-type and n-type thermoelectric legs, respectively. The cooling effect is opposed by Joule heating Q_j within the thermoelectric legs, found by:

$$Q_j = \left(\frac{I^2 (R_p + R_n)}{2}\right) \text{ measured in watts} \tag{19}$$

According to [8], it can be shown that half of the Joule heating passes to the sink T_c and half to the source T_h, with each half equal to (Eq. (19)). The expression for the heat energy absorbed at the hot junction, which is the same as the cooling power at the hot side of the thermoelectric couple Q_h can now be found by:

$$Q_h = (\alpha_p - \alpha_n) I T_h - (T_c - T_h)(K_p + K_n) - \frac{I^2 (R_p + R_n)}{2} \text{ measured in watts} \tag{20}$$

Hence, the efficiency of the thermoelectric couple η can now be found by (Eq. (16)), and written as:

$$\eta = \frac{P_L}{Q_h} \tag{21}$$

Substituting Q_h from (Eq. (20)) and P_L from (Eq. (14)) into (Eq. (21)), we obtain:

$$\eta = \frac{\left\{\frac{(\alpha_p - \alpha_n)(T_h - T_c)}{R_p + R_n + R_L}\right\}^2 R_L}{(\alpha_p - \alpha_n) I T_h - (T_c - T_h)(K_p + K_n) - \frac{I^2 (R_p + R_n)}{2}} \tag{22}$$

In thermoelectricity, efficiency is expressed as the dimensionless figure-of-merit Z, or more commonly, expressed as a function of the temperature over which the device is operated Z_T. The thermoelectric figure-of-merit Z describes the thermoelectric efficiency of a thermoelectric couple for a given pair of p-type and n-type thermoelectric materials, and is normally shown as:

$$Z = \frac{\alpha^2}{\lambda \rho} \tag{23}$$

where α is the Seebeck coefficient of the thermoelectric couple, λ is the thermal conductivity, and ρ is the couple's electrical resistivity. In the context of this discussion, the Seebeck coefficient of the couple is equal to $(\alpha_p - \alpha_n)$, the thermal conductivity λ is $(K_p + K_n)$, and the electrical resistivity ρ is equal to $(R_p + R_n)$, therefore (Eq. (23)) can be rewritten as:

$$Z = \frac{(\alpha_p - \alpha_n)^2}{(K_p + K_n)(R_p + R_n)} \tag{24}$$

The thermoelectric figure-of-merit Z is commonly used to optimize the performance of materials used in the manufacture of thermoelectric couples. However, when stating the thermoelectric efficiency, it is more common to express it as a function of the temperature over which the device is operated, referred to as Z_T, which can be found by using (Eq. (23)) at a specific temperature, and is normally written as:

$$Z_T = \frac{\alpha^2 \sigma}{\lambda} \tag{25}$$

where α is the Seebeck coefficient, σ is the electrical conductivity, and λ is the thermal conductivity of the couple at a specific operating temperature.

2.3 Thermoelectric module power generation

If we now consider a thermoelectric module that contains several thermoelectric couples connected electrically in series, and thermally in parallel, as shown in **Figure 2**, a small amount of electrical power, typically in the milliwatt range (mW), can be generated from a thermoelectric module if a temperature difference is maintained between two sides of the module. Normally, one side of the module is attached to a heat source and is referred to as the 'hot' side or T_H. The other side of the module is typically attached to a heat sink and is called the 'cold' side or T_C. The heat sink is used to create a temperature difference between the hot and cold sides of the module. If a resistive load R_L is connected across the module's output terminals, electrical power will be generated at the load when a temperature difference exists between the hot and cold sides of the module due to the Seebeck effect. A

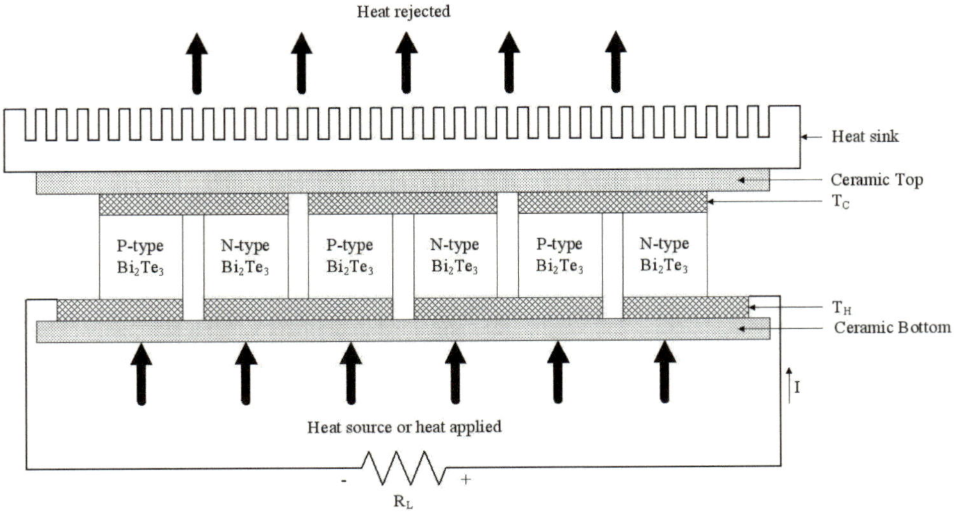

Figure 4.
A thermoelectric module configured for thermoelectric power generation.

schematic diagram of a thermoelectric module, operating as a thermoelectric power generator, is shown in **Figure 4**.

According to [11] the thermoelectric module open-circuit voltage V_{oc} can be obtained by:

$$V_{oc} = \alpha_m (T_h - T_c) \text{ measured in volts} \tag{26}$$

where α_m is the thermoelectric module's average Seebeck coefficient in volts per kelvin and may be found quoted in manufacturer's literature. If a load resistance R_L is now connected across the output terminals of the thermoelectric module, the current I can be found by:

$$I = \frac{V}{R_m + R_L} \text{ measured in amperes} \tag{27}$$

where V is the closed-circuit voltage and R_m is the thermoelectric module's average resistance in Ohms. As the Seebeck coefficient and module resistance are temperature dependent, their values should be calculated at the average or mean module temperature T_m given by:

$$T_m = \frac{T_h + T_c}{2} \text{ measured in kelvin} \tag{28}$$

The voltage at the load V_L can be found by:

$$V_L = I_L \times R_L \text{ measured in volts} \tag{29}$$

where the current at the load I_L is equal to the circuit current I. The electrical power generated by the module at the load can then be found by:

$$P_L = I_L^2 \times R_L \text{ measured in watts} \tag{30}$$

or by using:

$$P_L = V_L \times I_L \text{ measured in watts} \tag{31}$$

The efficiency η of the thermoelectric module can be found by:

$$\eta = \frac{P_L}{Q_h} \tag{32}$$

The cooling power at the hot side of the thermoelectric module Q_h can be found by:

$$Q_h = (\alpha_m) I T_h - (T_c - T_h)(K_m) - \frac{I^2 (R_m)}{2} \text{ measured in watts} \tag{33}$$

The efficiency η of the thermoelectric module can now be found by:

$$\eta = \frac{\left\{ \frac{(\alpha_m)(T_m)}{R_m + R_L} \right\}^2 R_L}{(\alpha_m) I T_h - (T_c - T_h)(K_m) - \frac{I^2 (R_m)}{2}} \tag{34}$$

The thermoelectric figure-of-merit Z_T can be found using:

$$Z_T = \frac{{\alpha_m}^2 \sigma_m}{\lambda_m} \tag{35}$$

where α_m is the module's average Seebeck coefficient, σ_m is the average electrical conductivity, and λ_m is the module average thermal conductivity, at a specific operating temperature.

2.4 A numerical analysis of a thermoelectric power generation module

If we consider a thermoelectric module, [12] have published some general material properties data for a 127 thermoelectric couple module. For the purpose of this analysis, we will use a hot side temperature T_H of 400 K, and a cold side temperature T_C of 300 K. The Seebeck coefficient and module resistance is temperature dependent, hence, their values should be calculated at the average module temperature T_m given by:

$$
\begin{aligned}
T_m &= \frac{T_h + T_c}{2} \text{ measured in kelvin} \\
T_m &= \frac{400 + 300}{2} \\
T_m &= 350 \text{ Kelvin}
\end{aligned}
\tag{36}
$$

According to [12], with an average module temperature of 350 K, the module thermoelectric parameters are; a Seebeck coefficient α_m of 0.05544 V/K; a module resistance R_M of 4.6491 Ohms; and a thermal conductivity K_M of 0.4422 W/K.

2.4.1 The thermoelectric module open-circuit voltage V_{oc}

The thermoelectric module open-circuit output voltage V_{oc} can be found by using (Eq. (26)) as:

$$
\begin{aligned}
V_{oc} = \alpha_m(T_h - T_c) = 0.05544 \times (400 - 300) \\
V_{oc} = 5.544 \text{ V}
\end{aligned}
$$

2.4.2 Connecting a load resistor R_L across the output terminals of the module

If we set the load resistance R_L to the value of the internal resistance of the thermoelectric module R_M of 4.6491 Ohms, the load resistance becomes:

$$R_L = 4.6491\Omega$$

2.4.3 The closed-circuit voltage V

The open-circuit voltage V_{oc} was calculated in 2.4.1, and in this case, when a load resistor R_L is connected across the output terminals of the thermoelectric module, the closed-circuit voltage V is:

$$V = 5.544 \text{ V}$$

2.4.4 The circuit current I

With a load resistance R_L connected across the output terminals of the module, the current I can be found by (Eq. (27)) as:

$$I = \frac{V}{R_M + R_L} = \frac{5.544}{4.6491 + 4.6491} = \frac{5.544}{9.2982}$$
$$I = 0.596 \text{ A}$$

2.4.5 The load current I_L and the voltage generated at the load V_L

The load current I_L is the same value as the circuit current I, hence:

$$I_L = 0.596 \text{ A}$$

The voltage generated at the load can be found by (Eq. (29)) as:

$$V_L = I_L \times R_L = 0.596 \times 4.6491$$

$$V_L = 2.7710\text{V}$$

2.4.6 The electrical power generated at the load P_L

The electrical power generated by the module at the load can now be found by (Eq. (30)) as:

$$P_L = I_L^2 \times R_L = 0.596^2 \times 4.6491$$

$$P_L = 1.6514 \text{ W}$$

or by using (Eq. (31)) as:

$$P_L = V_L \times I_L = 2.7710 \times 0.596$$

$$P_L = 1.6515\text{W}$$

2.4.7 The efficiency of the thermoelectric module η

The cooling power at the hot side of the thermoelectric couple Q_h can be found by (Eq. (33)) as:

$$Q_h = (\alpha_m) I T_h - (T_c - T_h)(K_m) - \frac{I^2(R_m)}{2}$$

$$Q_h = ((0.05544) \times 0.596 \times 400) - ((300 - 400)(0.4422)) - \frac{0.596^2(4.6491)}{2}$$

$$Q_h = 13.2169 - (-100 \times (0.4422)) - \frac{0.3552 \times (4.6491)}{2}$$

$$Q_h = 13.2169 + 44.22 - 0.8257$$
$$Q_h = 56.62\text{W}$$

Hence, the efficiency of the thermoelectric module η can now be found by Eq. (32) as:

$$\eta = \frac{P_L}{Q_h} = \frac{1.6515}{56.62}$$

T_H K	T_C K	Tm K	ΔT K	α_m V/K	K_m W/K	R_m Ω	R_L Ω	V_{oc} V	V_L V	I A	P_L W	Q_h W	Efficiency η %
300	300	300	0	0.05343	0.35	3.72	3.72	0.00	0.00	0.00	0.00	0.00	0.00
310	300	305	10	0.05343	0.35	3.72	3.72	0.53	0.27	0.07	0.02	4.65	0.41
320	300	310	20	0.05418	0.36	3.91	3.91	1.08	0.54	0.14	0.08	9.55	0.79
330	300	315	30	0.05418	0.36	3.91	3.91	1.63	0.81	0.21	0.17	14.40	1.17
340	300	320	40	0.05477	0.37	4.10	4.10	2.19	1.10	0.27	0.29	19.81	1.48
350	300	325	50	0.05477	0.37	4.10	4.10	2.74	1.37	0.33	0.46	24.90	1.84
360	300	330	60	0.05519	0.39	4.28	4.28	3.31	1.66	0.39	0.64	30.98	2.07
370	300	335	70	0.05519	0.39	4.28	4.28	3.86	1.93	0.45	0.87	36.33	2.40
380	300	340	80	0.05542	0.42	4.47	4.47	4.43	2.22	0.50	1.10	43.19	2.55
390	300	345	90	0.05542	0.42	4.47	4.47	4.99	2.49	0.56	1.39	48.82	2.85
400	300	350	100	0.05544	0.44	4.65	4.65	5.54	2.77	0.60	1.65	56.62	2.92

Table 1.
Summary of performance characteristics of a 127 couple thermoelectric module obtained by theoretical calculation.

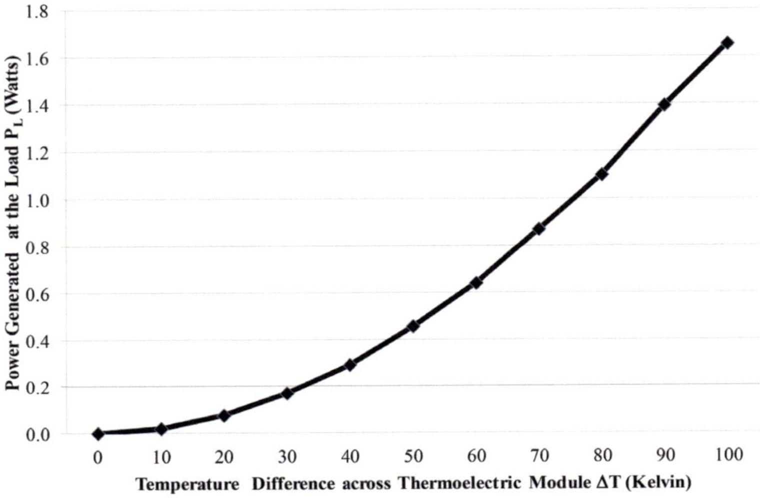

Figure 5.
Power generated at the load P_L with a temperature difference ΔT between 0 and 100 Kelvin (theoretical results).

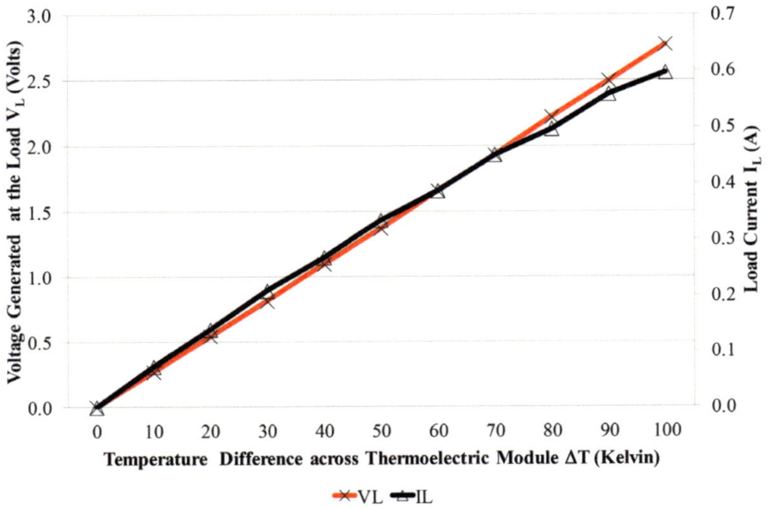

Figure 6.
Voltage V_L and Current I_L generated at the load with a temperature difference ΔT between 0 and 100 Kelvin (theoretical results).

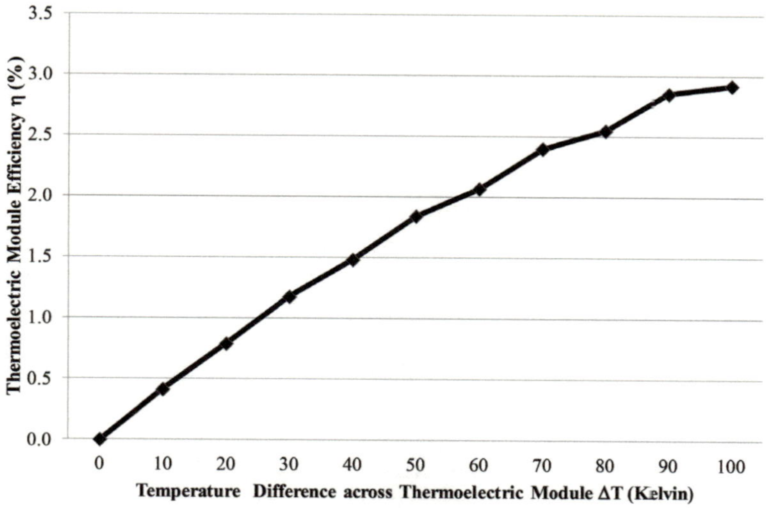

Figure 7.
Thermoelectric module efficiency with a temperature difference ΔT between 0 and 100 K (theoretical results).

$$\eta = 0.0292$$

$$\text{or } \eta = 0.0292 \times 100 = 2.92\%$$

2.4.8 Numerical analysis of the 127 couple thermoelectric module configured for thermoelectric power generation with a temperature difference ΔT varied between 0 and 100 K

A summary of the performance characteristics of the 127 couple thermoelectric module obtained by theoretical calculation is shown in **Table 1**. The temperature difference ΔT, which is equal to $T_H - T_C$, has been varied between 0 and 100 kelvin, in 10 kelvin steps, and the results calculated. The thermoelectric module's power generated at the load P_L, voltage V_L, current I_L, and efficiency η, is shown in **Figures 5–7** respectively. The results demonstrate the theoretical electrical power generated by a 127 couple thermoelectric module is typically in the mW to watt range when the module is subject to a temperature difference from 10 to 100 K.

3. Thermoelectric power generation and energy harvesting system

Thermoelectric power generation systems have typically needed to have a very high temperature gradient across the thermoelectric module(s) in order to achieve a useful electrical power output. This limitation has been a barrier to the successful application of this technology for power generation, and limited the technologies use to mainly niche applications, for example, in deep-space spacecraft power. However, with parallel developments in the area of electrical energy storage in supercapacitors, and low power DC to DC converters and boost converters, it is possible to develop a thermoelectric energy harvesting system that will operate from very low temperature gradients of around 1 K and be able to output useful power levels. This was previously very difficult to achieve and would have required several thermoelectric modules to be connected electrically in series, and thermally in parallel, increasing the overall system weight, size, and cost, and would only achieve relatively small levels of power generation unless a significant temperature gradient could be achieved across the modules.

The thermoelectric output voltage generated by a standard thermoelectric module can be boosted to a useful and stable level by using a low power boost converter and DC to DC converter. If the electrical power output from the DC to DC converter is then accumulated and stored for future use in a supercapacitor, it is possible to increase the potential output current of the system, and hence the overall power output of the thermoelectric energy harvesting system. A simplified block diagram of a thermoelectric energy harvesting system is shown in **Figure 8**, highlighting the five main stages of the system. The energy stored in the supercapacitor can be accumulated over time and released to the load when required. In some applications, it may be advantageous to use a voltage regulator after the supercapacitor in order to maintain a stable output voltage to the load.

In general, the duty cycle of the electrical load is a critical factor in determining the design of a thermoelectric energy harvesting system. As highlighted earlier, the output power of a single thermoelectric module is often too low to power other electrical and electronic components directly unless a significant temperature difference or gradient is available, or several thermoelectric modules are connected together electrically in series and thermally in parallel. The use of temporary electrical storage in supercapacitors leads to a focus on the duty cycle of the load, as it is necessary to ensure the capacitor can be recharged before the load becomes active to ensure repeatable and reliable operation.

3.1 Thermoelectric energy harvesting circuit

A thermoelectric energy harvesting circuit to power an electronic load is shown in **Figure 9**, and is based around a Linear Technology LTC3108 step-up DC to DC converter. The thermoelectric module output voltage is in the mV range when the

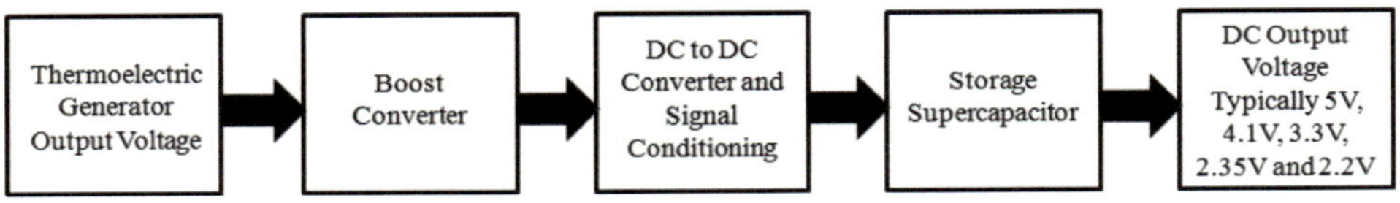

Figure 8.
Simplified thermoelectric energy harvesting block diagram [14].

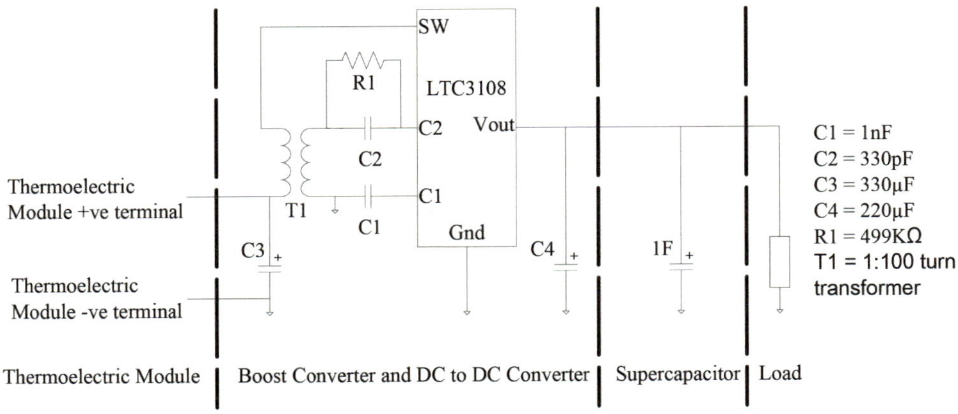

Figure 9.
Thermoelectric energy harvesting circuit [14].

module is subject to a small temperature difference, and is boosted to a useful level by the LTC3108 step-up converter. The LTC3108 uses a boost converter, in the form of an external step-up transformer, an internal MOSFET and associated circuitry within the DC to DC converter, to increase the voltage from the thermoelectric module. Within the converter, a MOSFET switch is used to form a resonant step-up oscillator using an external 1:100 turn transformer and a small coupling capacitor *C3* of around 330 μF. The frequency of oscillation is determined by the inductance of the transformer secondary winding and is typically in the range of 10–100 kHz. The AC voltage that is developed on the secondary winding of the transformer is boosted and rectified using an external charge pump capacitor *C1* of 1 nF and the internal rectifiers within the DC to DC converter. The DC to DC converter itself is powered via the internal VAUX circuitry, from the input voltage supplied by the thermoelectric module, and when the VAUX supply exceeds 2.5 V, the main output of the DC to DC converter V_{out} becomes operational and can be programmed by the user to one of four regulated voltages of; 2.35; 3.3; 4.1; and 5 V [13]. The converter operates at very low input voltages of 20 mV, which can be achieved when a 1 K or higher temperature difference exists between the 'hot' and 'cold' sides of the thermoelectric module. Dependent on the input power received from the thermoelectric module, the DC to DC converter output V_{out} will be charged over time up to its regulated voltage [13], and in this case, the 1F supercapacitor will charge up to 5 V at a maximum current of 4.5 mA.

3.2 Evaluation of the thermoelectric energy harvesting system

For the purpose of evaluation, a Melcor CP1.4-127-045L thermoelectric module and HX8-202-FM heat sink is connected for power generation, with a hot side temperature T_H of 323 K, and a temperature difference between both sides of the module ΔT of 9 K, the thermoelectric module's open-circuit voltage is 310 mV. An LTC3108EDE step-up DC to DC converter evaluation board, configured to output 5 V on the V_{out} pin, was connected to the thermoelectric module output terminals as shown in **Figure 9**. When the thermoelectric module is connected to the LTC3108 converter, the module's output voltage and current is 116 mV at 30 mA. The boost converter and DC to DC converter increase the thermoelectric output voltage and supply a stable voltage of 5 V at 0.4 mA to the 1F storage supercapacitor, charging the supercapacitor. The 1F supercapacitor reaches maximum charge after approximately 3 h [14]. Increasing the temperature difference across the thermoelectric module, and therefore increasing the thermoelectric module's output power, would have the effect of increasing the output current of the DC to DC converter, up to a maximum of 4 mA, reducing the capacitor charge time significantly.

If an electrical load is connected across the supercapacitor output terminals, the supercapacitor will discharge and supply electrical power to the load. For evaluation purposes, a number of different electrical loads have been tested, and successfully operate from the electrical power delivered from the supercapacitor during discharge including; a piezoelectric buzzer; light emitting diode (LED); humidity sensor; pressure sensor; low power microcontroller; DC motor; and a miniature electronic water pump.

The piezoelectric buzzer, LED, humidity sensor, and pressure sensor can be powered for a considerable amount of time before the 1F capacitor becomes discharged. For example, a Maplin electronic KU58 piezoelectric buzzer operates between 3 and 12 V with a maximum input current of 2.38 mA at 5 V, and successfully operates above 3 V for 18 minutes, and is still audible for in excess of one and a half hours when connected to the 1F supercapacitor. The miniature electronic water pump, DC motor, and low power microcontroller operate successfully from the 1F supercapacitor, however, the evaluation system would benefit from scaling the storage 1F capacitor appropriately to enable stable operation for significant periods of time. Notwithstanding this, and focusing on the miniature electronic water pump, an RS components 702-6894 electronic micropump that has an operating voltage of 1.2–6 V was connected to the output terminals of the of the 1F supercapacitor. The pump inlet and outlet tubes were connected together to form a 10 cm length of plastic tube, filled with water, and was seen to successfully pump this water around the tube for approximately 1 min and 30 s. Once the water pump minimum supply voltage of 1.2 V was reached during the first 35 s of the supercapacitor's discharge, the pump continued to operate, although at a reduced flow rate than observed earlier. The water pump draws a maximum input current of 72 mA at 5 V and can achieve a maximum flow rate of 150 ml per minute [14].

In general, a device with a; low input voltage; low current consumption; and a wide operating input voltage range; can be powered for a considerable amount of time directly from the supercapacitor. It should be noted that the performance of the two test circuits can be improved, with the addition of a resistor in series with the load in order to limit the current drawn by the load, or by using a voltage regulator after the supercapacitor, with an appropriate output capability, which would supply a stable output voltage to the load [14].

4. Discussion

The electrical output power obtained from a standard thermoelectric module is relatively small and in the mW range unless a significant temperature gradient can be achieved across two sides of the module, however, in common with other energy harvesting technologies, if the output voltage of the thermoelectric module is boosted to a useful level, i.e., to common battery supply voltages of; 5; 3.3; 1.5 V etc.; by using low power boost converters and DC to DC converters, and electrical energy storage in supercapacitors, practical thermoelectric energy harvesting systems can be realized which can output sufficient electrical power to operate low power electrical and electronic systems. This approach leads to a focus on the electrical load's duty cycle. Systems that rely on boost conversion and energy storage require the storage capacitor to be recharged in periods of load inactivity to enable continuous operation over an extended period of time. If the electrical load requires continuous power to operate, and the output power of the thermoelectric module is too low to power the load directly, the storage supercapacitor will eventually discharge completely and be unable to continuously power the load. However, in many applications the load does not need to be continuously powered—electronic sensors, microcontrollers, and RF networks often only need to make periodic readings, processing, and transmission cycles, and can be put into a low power 'sleep' mode when not in operation, drawing only a low quiescent current until 'waking-up' and drawing full current, allowing the supercapacitor time to recharge to a fully charged state. Applications where the electrical load is powered intermittently have become a focus for implementing low power thermoelectric energy harvesting systems.

The contemporary focus on low power energy harvesting systems will enable new thermoelectric applications to emerge and be realized. Thermoelectricity is commercially successful, having previously found applications in power generation for deep-space spacecraft power, military, and other niche applications. Recent focus has enabled thermoelectricity to replace or recharge batteries in low power electronic systems.

The short-term challenge for thermoelectric energy harvesting is to become a cost effective and practical solution to replace batteries in mainstream applications where access to temperature gradients or differences is available. Commercial success is often found in niche applications where there is a definite need and advantage, as the direct comparison with standard battery powered systems can highlight disadvantages of cost, size, weight, and complexity. However, as high volume thermoelectric applications emerge and are realized, the direct comparison with batteries will improve. Furthermore, with appropriate scaling and development, electrical rotating machines such as low power motors and pumps can be powered by a thermoelectric energy harvesting system, opening new applications and industries for exploitation.

The long-term challenge is to improve the efficiency and output power of thermoelectric modules, to develop new thermoelectric materials and module fabrication technologies, to reduce the cost of thermoelectric modules and energy harvesting systems, and to develop thermoelectric energy harvesting systems from low power to low-to-medium power applications.

5. Conclusions

Thermoelectric energy harvesting systems can be implemented to generate sufficient electrical power from naturally occurring or man-made heat sources to provide power to low power electrical and electronic components and systems. The duty cycle of the electrical load is, in general, a critical factor in determining the feasibility of implementation as, without the addition of power electronics and electrical energy storage, the output power of a single thermoelectric module is often too low to power other electrical and electronic components directly unless a significant temperature difference or gradient is available, or several thermoelectric modules are connected together electrically in series and thermally in parallel. To overcome this limitation, energy harvesting systems typically employ a low power boost converter and DC to DC converter to increase the thermoelectric module's output voltage to a useful level, i.e., 5, 3.3, or 1.5 V, and use a supercapacitor for temporary electrical energy storage and to provide power to an electrical load. The use of temporary storage in supercapacitors leads to a focus on the duty cycle of the load, as it is necessary to ensure the capacitor can be recharged before the load becomes active again to ensure repeatable and reliable operation. Successful applications have tended to be where there is a distinct need and advantage for implementation, and by replacing or recharging a battery in remote locations which are difficult to maintain and service. The short-term challenge for thermoelectric energy harvesting is to become a cost effective and practical solution to replace batteries in mainstream applications where access to temperature gradients or differences is available, and to be scaled to provide sufficient electrical power, and for the required duration, to enable electrical rotating machines such as low power motors and pumps to operate. The long-term challenge is to improve the efficiency and output power of thermoelectric modules, to develop new thermoelectric materials and fabrication technologies, to reduce the cost of thermoelectric modules and energy harvesting systems, and to develop thermoelectric energy harvesting systems from low power to low-to-medium power applications.

Author details

Chris Gould
Staffordshire University, Stoke-on-Trent, United Kingdom

*Address all correspondence to: c.a.gould@staffs.ac.uk

References

[1] Nuwayhid RY, Shihadeh A, Ghaddar N. Development and testing of a domestic woodstove thermoelectric generator with natural convection. Energy Conversion and Management. 2005;**46**:1631-1643

[2] Vining CB, Rowe DM, Stockholm J, Rao KR. History of the international thermoelectric society. In: Rowe DM, editor. Thermoelectrics Handbook—Macro to Nano. Boca Raton: CRC Taylor & Francis Group. 2006 Appendix 1-8

[3] Nolas GS, Sharp J, Goldsmid HJ. Thermoelectrics—Basic Principles and New Materials Developments. Berlin: Springer-Verlag; 2001. pp. 1-5

[4] Rowe DM. Thermoelectrics Handbook—Macro to Nano. Boca Raton: CRC Taylor & Francis Group. 2006. p. 4

[5] Riffat SB, Ma X. Thermoelectrics: A review of present and potential applications. Applied Thermal Engineering. 2003;**23**:913-915

[6] Bhandari CM. Thermoelectric transport theory. In: Rowe DM, editor. CRC Handbook of Thermoelectrics. Boca Raton: CRC Taylor and Francis Group. 2006. pp. 27-42. ISBN 978-0-8493-2264-8

[7] Rowe DM. General principles and basic considerations. In: Rowe DM, editor. Thermoelectrics Handbook: Macro to Nano. Boca Raton: CRC Press, Taylor & Francis Group. 2006. pp. 1-10-1-14. ISBN: 978-0-8493-2264-8

[8] Goldsmid HJ. Introduction to Thermoelectricity (Springer Series in Materials Science). Berlin: Springer-Verlag; 2010. ISBN: 978-3-642-00715-6

[9] LaBounty C, Shakouri A, Robinson G, Abraham P, Bowers J. Design of integrated thin film coolers. In: Proceedings of the 18th International Conference on Thermoelectrics (ICT1999); Aug 29th to Sep. 2nd 1999; Baltimore, USA. pp. 23-26

[10] Goldsmid HJ. Electronic Refrigeration. London: Pion Limited; 1986. ISBN: 0850861195

[11] Ferrotec. Thermoelectric Technical Reference—Power Generation. Ferrotec. http://www.ferrotec.com/technology/thermoelectric/thermalRef13/. [Accessed: 16th September 2013]

[12] Ferrotec. Thermoelectric Technical Reference—Power Generation. Ferrotec. http://thermal.ferrotec.com/technology/thermoelectric/thermalRefAA/ [Accessed: 16th September 2013]

[13] Linear Technology Corporation: LTC3108 Product Datasheet. LT0410, Revision A. Linear Technology Corporation; 2010

[14] Gould CA, Shammas NYA, Grainger S, Taylor I. Thermoelectric Power Generation—Properties, Application and Novel TCAD Simulation. In: 14th IEEE European Conference on Power Electronics and Applications (EPE2011); 30th August to 1st September 2011; UK: Birmingham

Chapter 3

Energy Harvester Based on Magnetomechanical Effect as a Power Source for Multi-node Wireless Network

Jerzy Kaleta, Rafał Mech and Przemysław Wiewiórski

Abstract

This work is focused on the development of new kind of energy harvesters that could be used in various applications including industrial, aerospace, or customer markets. The main aspect to consider is transformation of different sources of energy (that in normal conditions is wasted such as temperature, vibration, shock, etc.) into the usable electric power. The goal was to prepare wireless subsystem based on energy-harvesting technology which will aid different areas. The energy-harvesting devices are shown as small harvesting devices with power output from 10 mW up to 5 W. Proposed solutions might be used in applications such as low-power microprocessor systems, ultrasonic continuous power supply for low-power wireless network systems, and multi-node harvester systems that allow to collect more electrical power for critical structural health monitoring (SHM) applications. The main purpose was to obtain from harvesters the sufficient values for supplying the chosen 32-bit microcontroller systems. Additionally possible application in mechanic for the other than magneto-based solid harvesters is described.

Keywords: magnetomechanical cross effect, smart magnetic materials, magnetostriction, Terfenol-D, magnetostrictive actuators, frequency response, energy harvesting, harvesters

1. Introduction

The chapter describes the results obtained in the field of energy harvesting (hereinafter referred to as EH, also known in the literature as power harvesting or energy scavenging). EH is a set of methods that allow obtaining electricity from sur-rounding sources, such as mechanical, thermal, solar, and electromagnetic energy, salinity gradients, etc. [1, 2]. Energy harvesting is the use of sources commonly found in the environment (the so-called background energy), which are undesirable and usually suppressed (e.g., noise, shocks and mechanical vibrations of devices and structures, electromagnetic smog, heat as a result of friction and combustion, current flow, cooling engines, etc.) or widely available (sunlight, wave energy, salinity differences, biochemical processes, e.g., in plants), as well as those related to human biology (movement, body heat, etc.), for example [3]. Currently, it is assumed that EH can be an effective source of "cost-free" energy (after omitting the installation costs) for powering low-power devices (e.g. electronic devices, sensor systems, etc.). Hence the growing interest in civil and military applications. The area of use of EH concerns numerous civil and military applications and includes such disciplines as medicine, transport (cars, aviation, pipelines), construction structures (bridges, buildings), mechanical structures, sports and rescue equipment, and many more.

Energy harvesting creates new opportunities today, especially in the field of the so-called self-powered microsystems. This is a consequence of the progress in the field of materials and technologies enabling the recovery of energy from the so-called background, i.e., from known sources, but so far omitted, which in turn was due to the low efficiency of transforming energy and the high cost of producing the necessary devices for this purpose (the so-called harvesters). The decreasing energy consumption of these microsystems is also of key importance, which causes power sources with a power of a mile or even microwatts to acquire practical significance and allow to eliminate traditional power systems using cable systems or batteries or accumulators. A particularly promising area of EH applications is systems for continuous monitoring of inaccessible structures or biomedical implants, as well as distributed systems for the detection of threats on large surfaces (e.g., fire protection systems in forests or detection of chemical or radioactive contamination). It is predicted that in the near future the power of EH systems will increase significantly and will also have significance in industrial power engineering. A better solution is to take energy from the surroundings unlimited in time. It is assumed that in the future EH will be a source of high-power energy by creating appropriately extensive harvester networks.

The paper describes the main directions of EH research based on magnetic transducers and characterized numerous own constructions, including harvesters with magnetic processing using the Faraday effect and modal resonance, with a large increase in voltage under the influence of coil movement, with a moving core of austenitic steel and magnetostrictive core. In particular, the construction, selected characteristics, and possible areas of harvester use are described. The issue of miniaturization of the harvester's construction and modification of the magnetostrictive core was undertaken. Magnetographic field measurements were also carried out outside the harvester. In addition, harvesters using mechanical shock and a dedicated inverter as well as a low-power electronic system were presented. A method has been developed for the use of harvesters and actuators for the wireless transmission of energy and information using Smart Ultrasonic Resonant Power System (the so-called SURPS system), an autonomous system of diagnostics of environmental and operating parameters' multi-degree-of-freedom (Multi-DOF) and the so-called wireless harvesting nodes. The directions for further research have been defined at the end.

Authors recommend energy-harvesting solution to go (**Figure 1**), as an indispensable development system in EH applications. This is a versatile device from Würth Elektronik demonstrating the capabilities of EH and power supplies based on various sources of energy. There are two built-in subassemblies for obtaining energy: using a thermoelectric effect and photovoltaic panel. The electronic site consists of a series of inverters dedicated to EH by linear technology and EFM32 Giant Starter Kit (Silicone Labs) equipped in addition to the microcontroller including LED display and light sensor. This harvester provides the option of changing the configuration of connecting [4–6] power sources. This solution shows a multitude of potential power options using EH, which will be described later in the work.

1.1 Influence of smart materials on energy harvesting

Describing smart magnetic materials and taking into account their properties, it is difficult not to undertake in their own research the issues of their application in

Figure 1.
Energy-harvesting solution to go.

the field of energy recovery (EH) and—as discussed further—wireless transfer of energy and information. It can even be said that the development of EH is possible thanks to the advances in science and engineering in the field of smart materials, including those stimulated by a magnetic field. From a wide group of them, materials with a giant magnetostriction (GMM) were considered to be particularly worth the attention and acceptance as an object of research in the field of EH. The GMM properties are crucial here. A typical example is Terfenol-D. Materials with gigantic magnetostriction can convert magnetic energy into mechanical and vice versa. Thanks to such properties, these materials can be used in the construction of sensors, actuators, and harvester. GMMs obtain much larger deformations (Terfenol-D up to 70 times) than traditional magnetostrictive materials, and to achieve this effect, not very high magnetic field strength H is required. There is also the opposite effect.

Relatively small deformations generate relatively high magnetic field and therefore the induced electric current (in comparison with other ferromagnets). A very important feature of these materials is the wide range of operating temperature, as well as their low inertia (small hysteresis loop field), which facilitates their use in various conditions. The Curie temperature for Terfenol-D is 653–693 K, while the working temperature can reach up to 473 K. Examples of GMM applications in the EH range include aviation, road transport, stationary mechanical structures, medicine, sports and tourism equipment, and many more. The aim of the research is mainly to increase the efficiency of converting mechanical energy into electricity, miniaturization of harvesters, and reduction of their price. Solid Terfenol-D, despite its many advantages, has several disadvantages that hinder its wider application in the field of EH. A significant drawback is, above all, the high brittleness, which is associated with low tensile strength. Another limitation is eddy currents of considerable value, which limits the effective frequency of operation of the devices to several kilohertz. An important parameter is also the price of Terfenol-D, which remains at the level of 1 $/1 g. These disadvantages are the reason for searching for new solutions. One of them is magnetostrictive composites, which can also be used in the construction of harvester.

1.2 Wireless energy and information transfer through energy harvesting

The use of harvesters increasingly requires solving the problem of unconventional energy and information transfer by solid, liquid, and gas media. Therefore, it was considered important to characterize the state of the art in this area and undertake own research. Smart materials in this case can be effectively used for wireless energy and information transfer using ultrasonic vibrations. Most often, piezoelectric and magnetostriction transducers are used for this purpose. There are many ways of wireless power transmission (WPT) using various couplings, e.g., inductive (most popular today since the pioneering work of N. Tesla), capacitive coupling, microwaves, optical coupling, and sound waves, including ultrasound. This last opportunity has been known for over 40 years. In 1970, the first paper [7] appeared, indicating the possibility of using ultrasounds not only for medical or engineering research but also as an energy carrier in transmission through solid bodies. The ease with which ultrasound passes through the solids was then observed. In 1998, using the given idea, a special heart stimulation electrode was patented for arrhythmia [8]. The biomedical application of ultrasound for energy transmission is intensively developed. Particularly noteworthy here is, for example, work [9], which shows the way of powering, using ultrasound, an actuator placed in the human body. The idea of this type of power lies in the fact that in the receiver, the energy of ultrasonic waves and vibrations caused by them are not converted back to the electrical voltage at all, but through the arrangement of vibrating elements—they directly supply the actuator. Other interesting studies in this area are described in [10, 11].

Much attention is devoted to energy conversion efficiency using piezoelectric transducers. The most frequently cited are works [12–14]. The last of them showed efficiency at the level of 50% in the transmission of ultrasound in the air at a distance of 70–80 m. Equally spectacular achievements are the Dutch team [15, 16]. Another interesting example is the transmission of data with the help of ultrasound over the plane [17, 18]. It should be noted that very intensively developed activities are aimed at mastering the effective transfer of energy and information through thick metal barriers, mainly using piezoelectric harvester. These works, initiated in the United States by Saulnier in 2006 [19], gained interest in the navy due to the possibility of sending energy and information through the thick walls of submarines. Particularly significant results were published in the dissertation of Lawry [20] and in a dozen or so publications after its defense, e.g., [21, 22]. The last two indicate that the state of knowledge allows the use of such relays on submarines today. The continuous supply of approximately 50 W of electricity, along with 12.4 Mbps of data through 2.5-inch (over 6 cm) metal walls, is an ideal system for use in submarines that require avoiding leakage and high safety. In [23], it is also pointed out that the system can also be used in ships, unmanned vehicles, armored personnel carriers, tanks, and airplanes.

Another large American project funded by the National Aeronautics and Space Administration (NASA) is research conducted by the team of Sherrit from the Jet Propulsion Laboratory. The research, begun already in 1998, concerned the possibility of generating and reading ultrasound signals using piezoelectric actuators and generators [23]. In 2005–2008 this technique was constantly improved. In 2005, the theoretical basis for energy transmission by flexible materials of thickness over 1.5 cm [24] using piezoelectric actuators was described and then—to improve efficiency—using special graphite "patches" attached with thin layers on both sides of the wall [25]. Obtained results were promising, and it was decided to do the first trials with use of mentioned above technology in the vehicles of the NASA. The team's many years of work have been summarized in a comprehensive publication on the physical basis of ultrasonic harvesting [26] with the use of piezoelectric receivers and transmitters. Recently, a team led by Sherrit has developed a method for feeding the stepper motor through the metal wall of the vessel [27]. Thanks to the uniform power transmission, it is possible to continuously control the motor by the generated ultrasonic waves. Interestingly, this wave is not converted here to electricity and again to the mechanical energy of the engine, but the vibrating elements cause the motor to move directly by picking up ultrasonic waves. A broader literature analysis in the field of power transfer using ultrasound was carried out in [28].

2. Review of EH methods capable of supplying wireless harvesting nodes

Currently, there is a trend to create autonomous power supply systems for low-power consumer electronic devices (including the so-called toward zero-power information and communication technology (ICT)) or a variety of sensor systems and monitoring systems (e.g., structural health monitoring (SHM)), e.g., [6]. It is assumed that even the use of batteries in these cases is not an optimal solution, e.g., due to the troublesome replacement of batteries and their recycling.

The development of the physics of cross-field phenomena, in which one field (e.g., mechanical, thermal, magnetic) enables energy to be obtained in a different form (e.g., electricity), progresses very quickly and is supported by achievements in the field of material engineering. This results in the fact that there are an increasing number of materials usually called smart, which can be effectively used to build harvester.

The number of physical phenomena that produce electric current is significant, e.g., [29–32]. You can include here:

- Piezoelectric effect
- Reverse magnetostriction (Villari effect)
- Faraday electromagnetic induction phenomenon
- Thermoelectric effect (Seebeck effect)
- Static electricity
- Differences in superconductor parameters
- Pyroelectric effect
- Ionization using an electromagnetic field

EH can also be realized using double cross fields, for example, first heat and then electric current. The interdisciplinary nature of the issue, which is energy harvesting (physics of cross effects, material engineering, mechanics, electronics), stimulates the development of science and the economy. It should be emphasized that, despite numerous works undertaken mainly in the last decade in the research centers of the most developed countries, the subject of EH and the various smart materials used for this purpose is still very topical in terms of science and application. Leading economies and research centers allocate significant resources to basic and applied research in the field of EH.

Due to scientific goals and interests, further work was focused on the use of methods increasing the parameters of harvester, mainly energy conversion efficiency, using the acquired experience in the field of magnetomechanical cross effects, smart materials, strength of materials and mechanical structures, and measurement methods. The extent of the subject matter required the imposition of restrictions. Therefore, magnetostrictive harvesters using GMM-type materials were recognized as key. Thanks to their application, instruments that were able to recover energy from sources not yet explored such as mechanical impact were obtained. An important limitation of the magnetic core harvester is its size and weight. Installing piezoelectric harvesters is a lot simpler than a magnetic core harvester that requires a complicated mechanical construction, premagnetization,

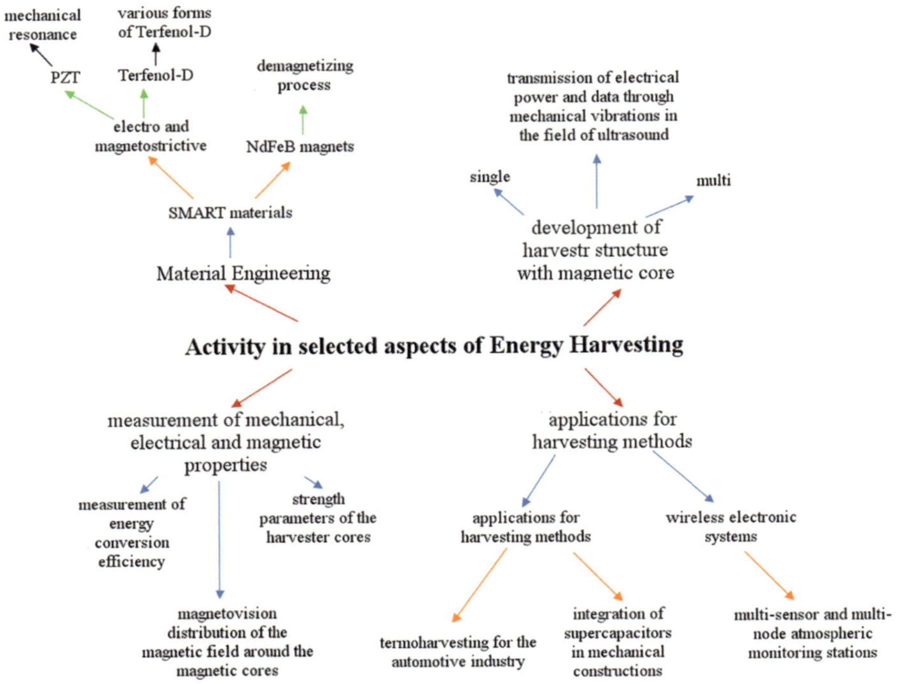

Figure 2.
Determining the dominant issues taken from energy harvesting.

and prestress. However, the current-voltage performance of magnetostrictive harvester is an order of magnitude larger than other types of harvester. That is why this type of harvester was considered to be particularly interesting. Further designs will include further miniaturization of the instruments. Current and future interests in this area are characterized by the graph presented in **Figure 2**. The following examples of implemented harvesting technologies are only briefly presented, the broader discussion of which will take place later.

In the field of low-power technique, the definition of harvester as the power supply of a single microprocessor was adopted, which after powering wirelessly sends data in accordance with its operating algorithm (program code) to the receiving and processing unit. A single harvesting system is a node in a larger structure managed from a central site. Individual configurations of harvester can make it easier to tune the harvesting power for specific phenomena that trigger its operation.

2.1 Harvester as an electric generator

The obtained power in laboratory harvesters became bigger; hence these devices were treated as (source) an electric generator. Due to the physical phenomenon used for the EH effect, construction, the principle of work, the conditions in which the harvester works, and characteristics of the source, harvesters can be divided, as follows:

- Constant voltage (e.g., harvesters based on a thermoelectric effect)
- Variable voltage (e.g., harvesters based on the Faraday effect, e.g., as Piezo patch)
- Impulse (e.g., solid-state harvesters, e.g., top core coil magnet (TCCM)

The pulse supply differs from the voltage-variable frequency of the occurrence of force and the instantaneous value of the generated current. Voltage supply is characterized by frequencies similar to the electricity in the electrical network (50/60 Hz). The generation of the voltage in the impulse supply occurs rarely and for a very short time, but the amplitude is very large. Due to the characteristics of the harvester circuits, they can be divided into:

- Current sources (Faraday generator, magnetostrictive harvesters)
- Voltage sources (Piezo patch type)

2.2 Types of electrical circuits due to the type of energy source

The essence of EH is to create new concepts of current generators, using cross effects, including more often magnetomechanical phenomena. It is assumed that

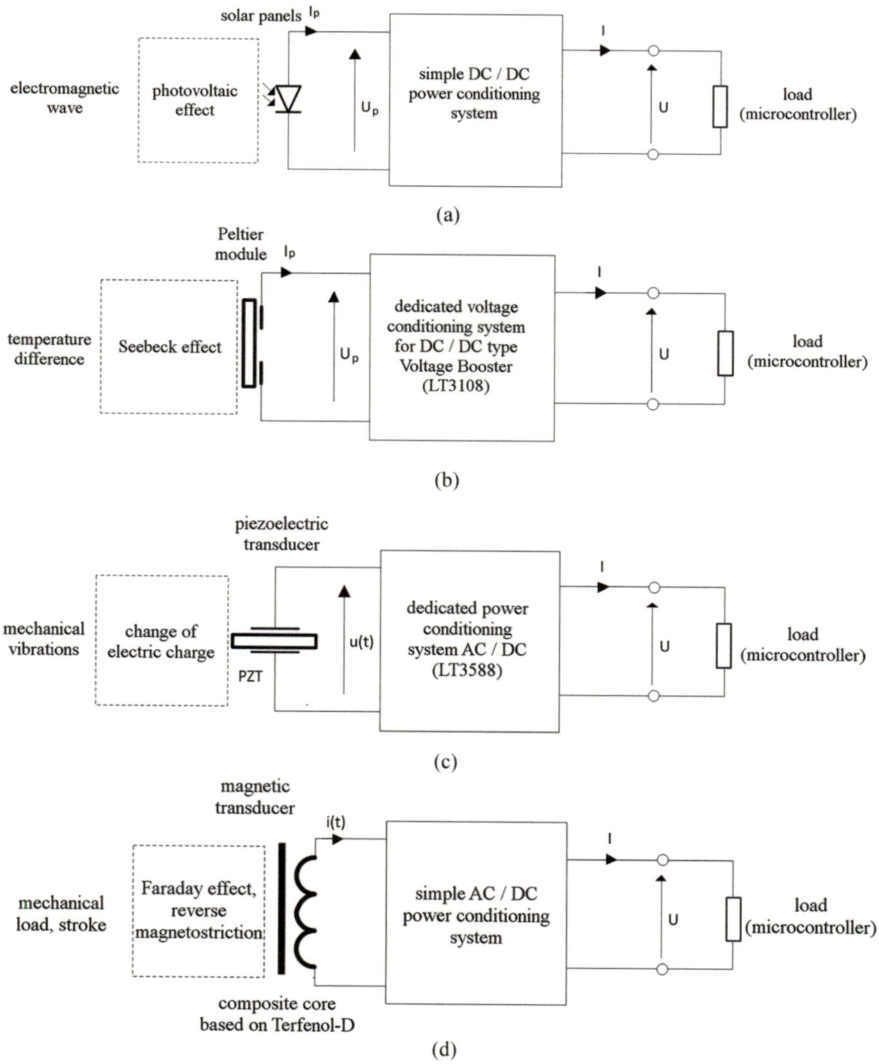

Figure 3.
Configurations of electrical circuits due to energy recovery from a specific source and converter: (a) solar, (b) temperature differences, (c) piezoelectric transducer, and (d) magnetic transducer.

even for small power and efficiency, it can be a valuable power source. The development of the technology of constructing harvester, with similar electrical parameters as chemical cells, may reduce the production of the latter for ecological reasons. As harvesters acquire energy in a nonparasitic manner, i.e., they process energy considered as a by-product ("junk") process, they increase the efficiency of the system as a whole.

Both electricity and electric voltage must have the parameters necessary to supply both the sensors themselves and the built-in processor with the transmitter adapted to it, as well as the communication unit. Another problem is the conversion and conditioning of the voltage/current from the generator (**Figure 3**) [33, 34]. Designing electrical circuits for harvester requires knowledge of the device's operating characteristics.

Only harvesters based on a thermoelectric or photovoltaic effect generate DC current. Harvesters recovering energy from vibration, magnetostrictive, and piezoelectric, as well as based on the Faraday effect, are on the other hand alternating current sources. Harvesters powered by impact impulse are a special case [35]. The generation of electricity in pulsed power takes place for a very short time, but the current amplitude is very high. Harvesters "powered" by mechanical shock generate a variable voltage waveform and are characterized by a strong current pulse, and in the generated signal, there are frequencies related to magnetic resonance of the core-coil system magnetostrictive core.

3. Magnetic-based effects of solid-state energy harvesters

Electricity can be generated by operating on a coil with a variable magnetic field. Such a field can be induced by another coil, in which a variable current flows, we are talking about the mutual induction of coils. This is how the transformer works. By definition, a harvester should be designed so that it does not require additional power supply. Materials that can be used to generate a variable magnetic field are:

- Permanent magnets (e.g., neodymium NdFeB), which are a source of constant magnetic field. In order to be able to recover energy through a coil, a source of an alternating magnetic field is necessary, which means movement of the magnet-coil system against one another.

- Materials with gigantic magnetostriction (giant magnetostrictive material—GMM): Work on new materials has led to the development of materials with gigantic magnetostriction, which undergoes the action of force, deforming, while generating a variable magnetic field. If harvesters based on these methods are subjected to mechanical vibrations, which are a side effect of a certain process, they can be considered a "free" source of alternating electric current, resulting from the appearance of a variable magnetic field generated in the coil, obtaining the best energy conversion parameters [36–40].

As part of our own research, we selected a group of smart materials for harvesting applications and developed many solutions and harvesting methods predestined for the SHM [28, 41] application. The scope of works on magnetic harvester is presented in **Figure 4** [42]. Harvesters with a smart magnetic core can be used as:

- Impulse power supply operating under the influence of mechanical impact with energy adjusted to the size of the harvester core, conditioning its electrical power

- As an electric power transmitter operating under the influence of ultrasonic vibrations above 25 kHz, supplied either from an actuator or a specific technological process

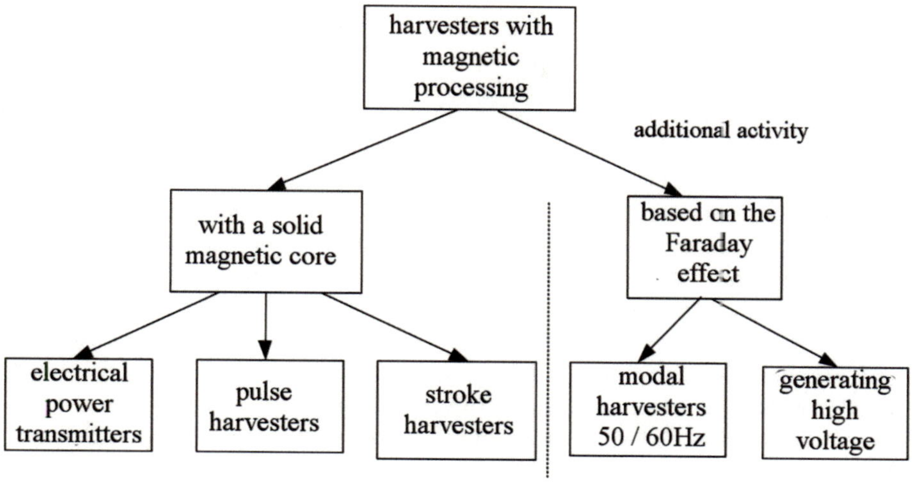

Figure 4.
Types of harvesters with magnetic processing.

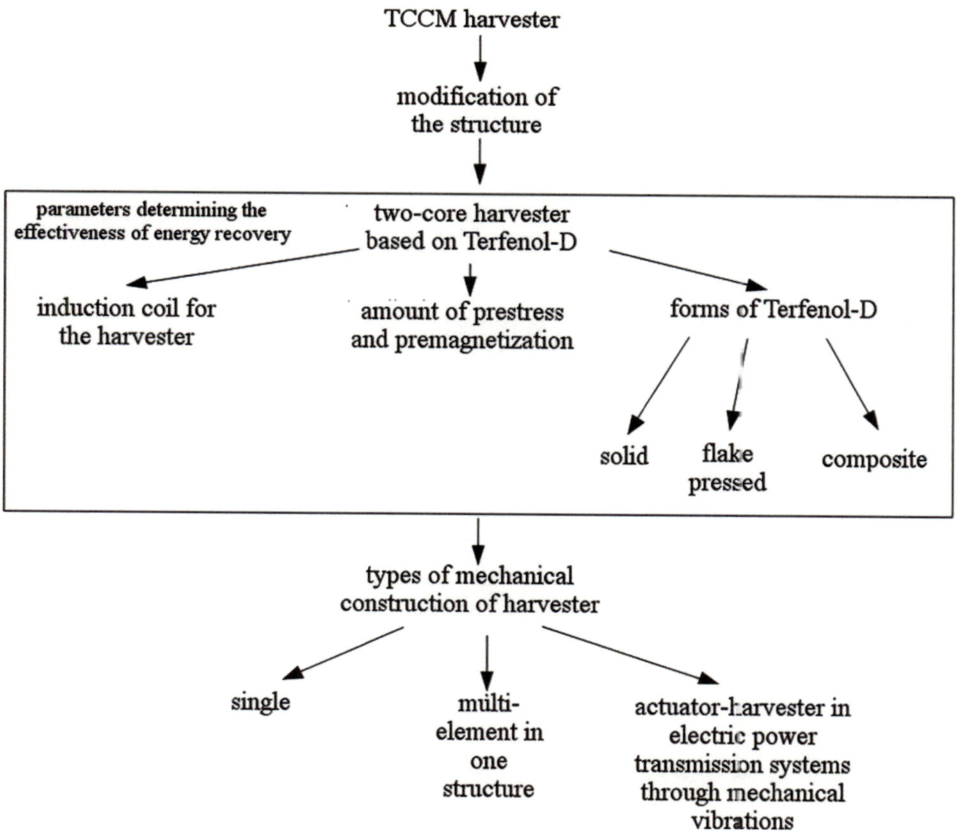

Figure 5.
Development of the next generation of harvester with magnetic core and their applications.

3.1 Harvester construction: core modification

In the next step, it was considered advisable to undertake the task of miniaturizing the harvester structure by modifying the harvester core. The magnetic circuit of the core consists of a set of permanent magnets coupled with magnetostrictive elements, which in turn are cores made of Terfenol-D, solid, as well as in the form of compressed flakes, which makes it possible to reduce eddy currents. Proper selection of parameters related to prestress and magnetization of the magnetostrictive material ensured the supply of the microprocessor even with much smaller dimensions than the one described in [33]. In addition to the typical design assumptions, it was necessary to formulate assumptions from the electrical and functional side, which would ensure a total possibility of working in the mode of actuator-harvester [33].

Previously, harvesters with magnetic processing have been described. The subgroup of magnetic harvester is top core coil magnet (TCCM) harvesters and its variants, double top core coil magnet (DTCCM) and TCCM model 2 [36]. In the work [36], it was shown that harvesters have low electric power yields in relation to dimensions and weight. Further work was aimed at developing a new harvester structure capable of miniaturizing the device without compromising performance and electrical efficiency. The schedule of work on the development of the structure is presented in **Figure 5**.

In order to develop a miniature harvester, the following assumptions were made:

1. As a core solid Terfenol-D must be used.

2. The core of Terfenol-D will be wrapped with foil, which will protect it against crumbling.

3. Alignment will be followed by a "cone-hole" pair.

4. An NdFeB magnet will be placed inside the coil, which will increase the obtained results with the Faraday effect under the influence of core magnetostriction.

Figure 6 shows a comparison of the TCCM structure currently developed based on two solid Terfenol-D cores. The fact that two external NdFeB magnets have been placed inside nonmagnetic oscillating cones is noteworthy.

The prestress of the core is determined by tightening the thread between the clamp and the body. A hole has been made in the aluminum cover in which a cylindrical-shaped ring is received, which receives vibrations. Between the clamp and the washer, there is a rubber ring, which acts as a shock absorber for transmitted vibrations and determines the prestress. By changing the mutual position of the clamp and body, the force at which the polyurethane ring is compressed is influenced, which acts on the core-coupled washer as shown in **Figure 7**.

The harvester body is made of steel; it acts as a seismic mass, affecting the core through a cone embedded in the hole at its bottom. A hole has been made in the body with a diameter suitable for the coil with magnetostrictive cores and a supply opening for the coil wires. The advantage of the body is that it shields the magnetic field from the magnetic circuit. All elements and assembly of the harvester are shown in **Figure 7**.

3.2 Review of prototypes of harvester with a smart magnetostrictive core

The above described only selected own works which allowed to create a palette of harvesters. The type of work and power range of the harvester are shown in the graph below (**Figure 8**):

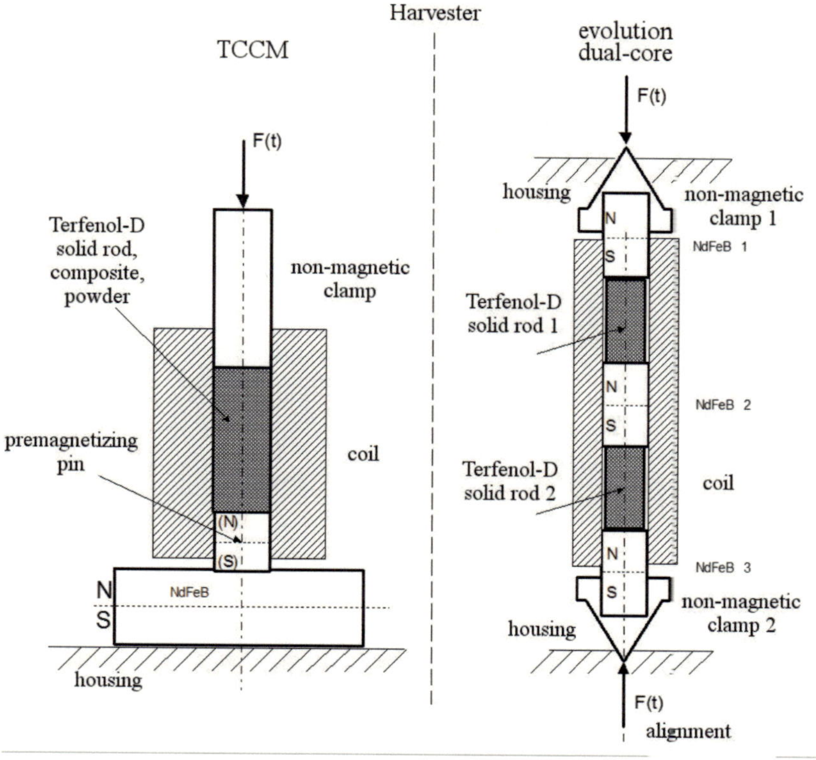

Figure 6.
Comparison of the structure of harvester construction developed in the Laboratory of Dynamics of WRUT.

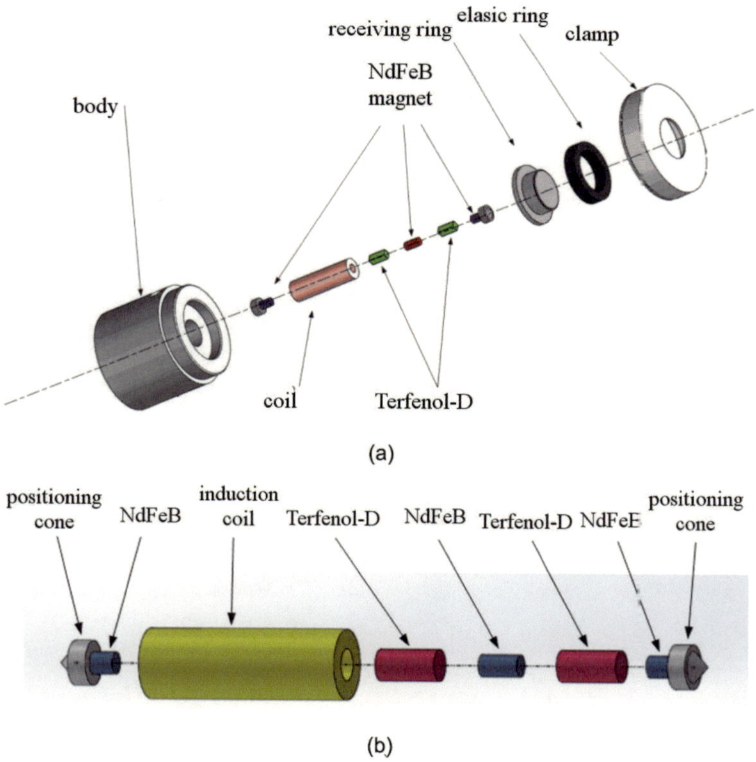

Figure 7.
List of all elements of the harvester: (a) main body, (b) coil with inner elements.

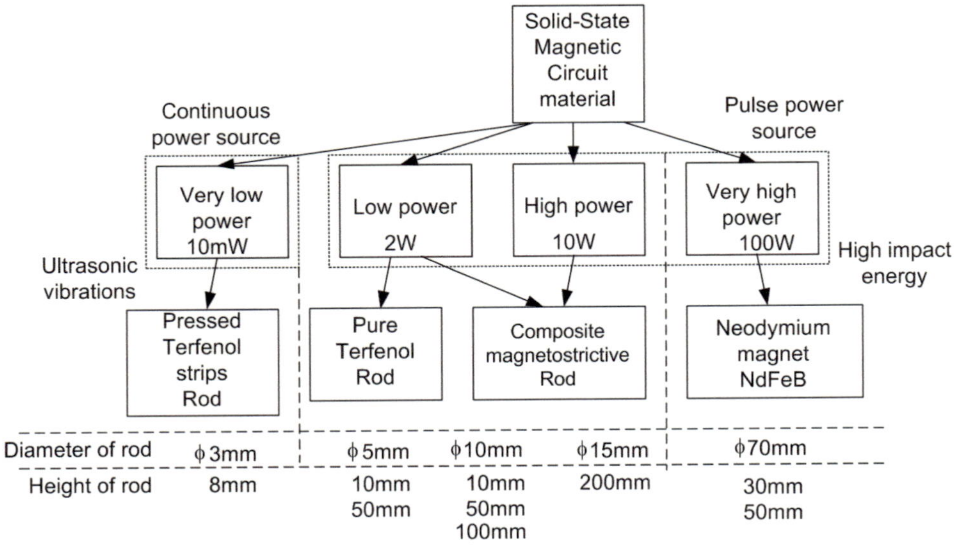

Figure 8.
The classification of harvesters due to the nature of work, power, and the core material.

Figure 9.
A view of prototypes developed in the Laboratory of Dynamics WRUT (a) and (d) low power harvesters, (b) high power harvester, (c) "Tactical grade" type harvester.

The electric power obtained by harvesters depends on the type of material and dimensions of the core. A separate group consists of solutions based on the method of

Harvester type	View from Figure 9	Dimensions [mm]	Mass [g]	Power in the impulse [mW]
Miniature	—	ϕ50 × 35	200	2000
Low power	(A, D)	ϕ50 × 50	300	5000
High power	(B)	ϕ50 × 150	1000	10,000
"Tactical grade"	(C)	ϕ80 × 250	1500	10,000

Table 1.
An overview of the prototypes of harvesters together with the most important parameters.

demagnetizing magnets, in which neodymium magnets subjected to strong mechanical stimulation, sometimes to its destruction, are used as the core. They differ in the amount of Terfenol-D used in the magnetic circuit and provide the possibility of screwing into the structure. The view of made harvesters is shown in **Figure 9** (**Table 1**).

4. Harvesters based on mechanical impact and electronic converters for small power applications dedicated for them

Generators of special characteristics are the explosive-driven ferromagnetic generators (EDFMG) producing an electromagnetic wave that arises as a result of immediate demagnetization of the magnet by a stroke following an explosion or other strong force impulse. The magnet then loses its magnetic properties but generates a strong pulsed magnetic field around it. During the impact it is possible to even destroy the magnet, but the amount of energy that will be induced in the coil is large, and it is enough to charge high-voltage capacitors with a large capacity. This issue is the subject of intense research, especially in the last decade, and their goal is applications, mainly military [43, 44].

One of the proposed methods of generating electricity directly from the impact was the impact demagnetization of NdFeB permanent magnets [35, 36]. Just as a spring has its constant, which is a measure of energy accumulated in it, the magnet has similar storage properties. Large diameter springs have large solid, strong magnets and have a high energy density. Permanent magnets containing components of rare earth have the highest energy density (see **Table 2**). This applies to the generation of electricity for the instantaneous supply of microprocessor systems from the impact demagnetization of permanent magnet-type NdFeB. Currently, NdFeB magnets are the most powerful permanent magnets. The advantages of NdFeB permanent magnets in impact harvesting:

- The largest—of all permanent magnets—BH energy, up to 600 kJ/m^3
- Strong magnetic flux at a surface of up to 2 T
- High hardness of the structure with simultaneous resistance to cracking

The disadvantages of neodymium magnets include:

- Poor resistance to thermal changes—high temperature has a destructive effect on the BH parameter.
- Oxidation of the outer layer of the magnet makes it necessary to use chromium as the outer layer.

Material	Energy density [kJ/m³]	Bhmax MGsOe	Remanence kGs	Coercion kOe
N27	199–223	25–28	10.2–11.0	Min. 9.6
N30	223–247	28–31	10.8–11.5	Min. 10.0
N35	263–286	33–36	11.7–12.1	Min. 10.9
N38	286–302	36–38	12.1–12.5	Min. 11.3
N40	302–326	38–41	12.5–12.8	Min. 11.6
N42	318–342	40–43	12.8–13.2	Min. 11.6
N45	342–366	43–46	13.2–13.8	Min. 11.0

Table 2.
Magnetic parameters of permanent magnets NdFeB divided into classes.

Through the use of NdFeB magnets, harvesters feature the smallest possible external dimensions. The process of releasing energy through a surge load has become a determinant for the construction of a new generation harvester.

The limited availability and increasing price of rare earth elements are the reason for reducing their share in the composition of permanent magnets. At the same time, this is the reason for intensive research on improving the operational performance of magnets, with a significant reduction in manufacturing costs. On the other hand, you can see that, despite the popularity of the so-called neodymium magnets, not all of their capabilities have been noticed and fully used. There is little work on the use of NdFeB magnets as energy storage sources, which are used if necessary due to demagnetization as a result of mechanical impact. Due to the "longevity" of magnets, you can store "programmed energy" in them much longer than in typical alkaline batteries or accumulators. Of course, the amount of stored energy is much smaller than in typical lithium batteries, but in the case of energy recovered from magnets, there are no limitations in the type of leakage current, causing self-exhaustion of batteries. It is also possible to recycle magnets after fully demagnetizing them. Assumptions adopted during the construction of the impact harvester:

- Neodymium magnet with "stored" energy (BH) max can be treated as a warehouse with energy that can be used with impact demagnetization.
- The harvester can be stored in conditions much less favorable than typical batteries, even in seawater.
- The visible trend of reducing rare earth elements will result in a decrease in the cost of producing the harvester; however, a NdFeB magnet should be used as the method of standard.
- In the magnetic circuit, the simplest construction of the magnetomechanical magnetic circuit should be used (application of pre-pressure, magnetic screens).
- Energy "recovered" from a permanent magnet can be used to power a low-power sensor system.
- Electronics used in the input stage supplying microprocessor elements should have a minimum starting voltage of several mV.

There are examples of ferromagnetic generators which, thanks to the impact (explosion) against the neodymium magnet, obtain instantaneous powers reaching MW; however, in the energy-harvesting application, there is no need to destroy the magnet but only a "light" impact that would not cause its rapid destruction. The way to convert energy into electricity is to place the magnet in the induction coil, just as it is placed in it and other materials, e.g., in electromagnets or in the case of Terfenol-D [2]. Due to the wave phenomena resulting from the stroke, the coil must have a special construction, also due to the polarity of the NdFeB magnet. In the case of a wave transition, a large number of windings are not required (**Figure 10**).

Due to the pulsed energy release, too high inductance of the magnetic circuit causes the reduction of the recovered current due to the increase of the substitute output impedance. The winding should be permanently attached to the magnet. One of the most important information about a magnet that cannot be omitted is the shape and arrangement of the zero line. The winding should be made only at one of the poles, N or S. This means that the magnet should have the largest possible height to diameter ratio but at $ > 10 mm. Currently a 0.6 ratio is assumed to be the standard; however, there are solutions with a proportion close to 1. Be careful about the arrangement of the zero line, which shifts under the influence of demagnetization, and do not combine magnets in NSNS cascades, because the resulting relaxation of the NS transition results in a dramatic reduction in the performance of the magnetic circuit. A good chance to improve the performance of the recovered current is to use the Halbach matrix as the object to be demagnetized.

Harvesters using the mechanical impact phenomenon generate a variable voltage waveform. At the same time, it is characterized by a strong current impulse, and in the generated signal, there are frequencies associated with magnetic resonance of the core-coil system. Next, a new method of acquiring electric current is presented as a result of demagnetizing neodymium magnets in a circuit with a magnetostrictive core.

4.1 A dedicated low-power electronic system for impact harvesters

The use of a small number of coils around the magnet enables the "capture" of rapid change of the magnetic flux and the generation of electricity directly from the magnet impact. However, a very low voltage level at a very high current requires the use of specialized electronic transducers capable of delivering the

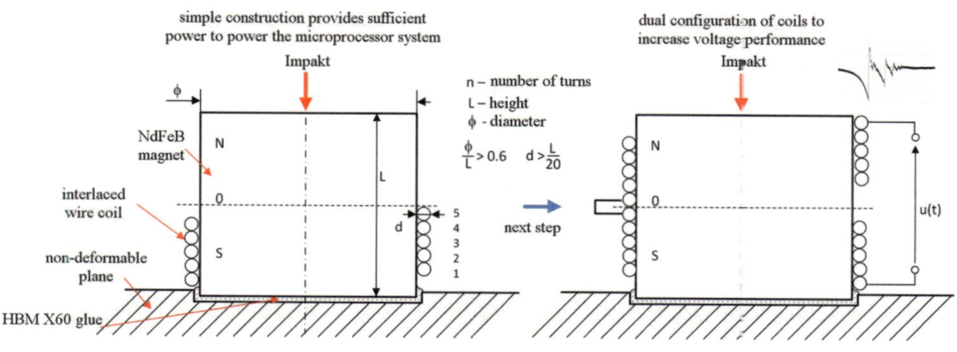

Figure 10.
The scheme of MFT harvester construction together with the description of the relevant parameters.

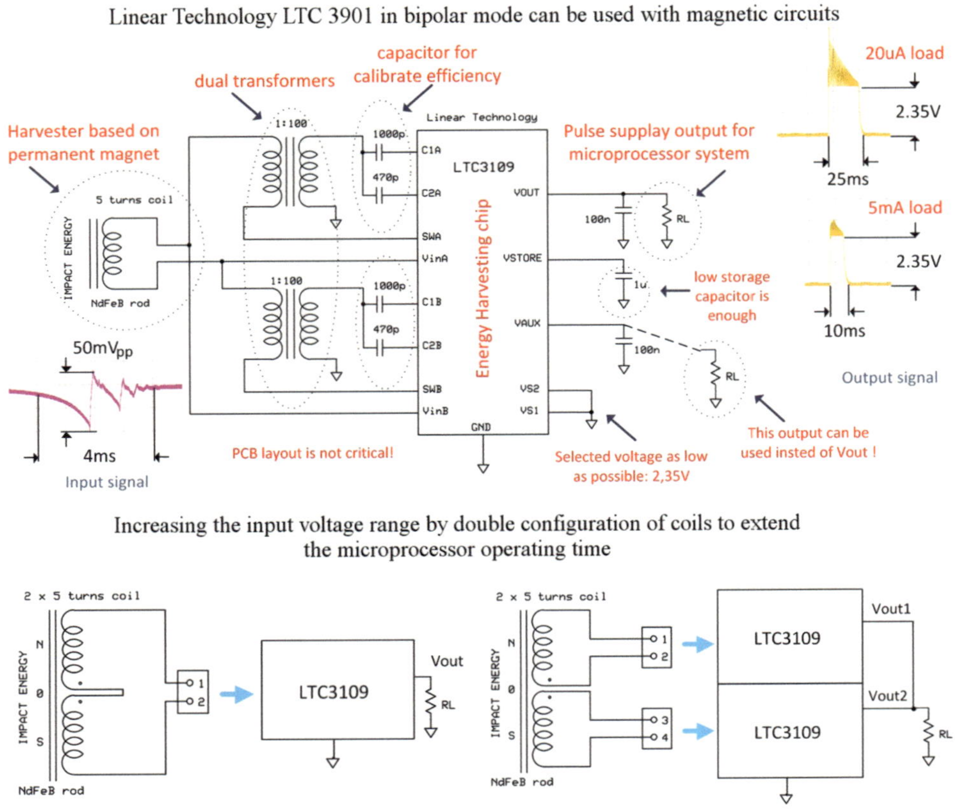

Figure 11.
Application of linear technology LTC3109 as a power conditioner for the microprocessor.

appropriate voltage level to power the microprocessor system. The linear technology LTC3109 system dedicated to thermoelectric applications operating in a bipolar configuration was used in an original way, which, as it turned out, enables voltage processing from low-impedance magnetic circuits. The obtained results demonstrated the usefulness of the system to resonant frequencies close to 70 kHz. The most important features of the harvester with the LTC3109 system are shown in **Figure 11**.

Thanks to the LTC3109 system, which enables the capacitor to be charged for the shortest possible time (the dynamic resistance parameter (ESR)), the operating time of the microprocessor is extended. It is estimated that the harvester subjected to a stroke with a 1 ms force impulse at the energy storage capacitor 100 nF manages to extend the microprocessor operating time to 6 ms. This effect is presented in **Figure 12**, and the view of the prototype impacts harvester system on **Figure 13**.

It should be borne in mind that the estimated efficiency of transforming the impact of the magnetization of the neodymium magnet to electric current is only 0.02%. Therefore, the key challenge is better transformation of energy, which requires changes in the harvester construction. An important aspect is the standardization of the harvester, in terms of their geometric dimensions, conditioned by the application requirements. It is possible to create their series of types, from miniature versions to powers of several watts, as well as relatively large ones (e.g., with neodymium magnets Ø100 mm diameter) for applications in, for example, mining. Further works should also consider the possibility of replacing relatively expensive neodymium magnets with their counterparts lacking rare earth elements.

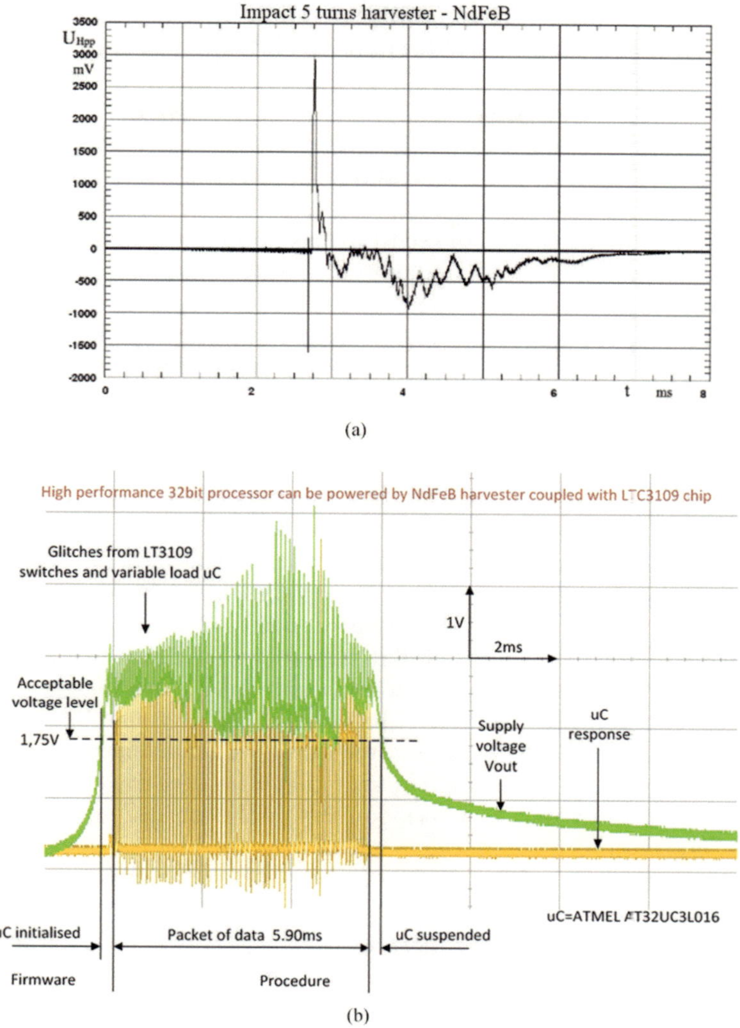

Figure 12.
(a) An example diagram obtained as a result of stimulation of a harvester by mechanical shock, (b) waveforms on the conditioner LTC3109, and microprocessor by Atmel together with a description of the parameters.

Figure 13.
View of the impact harvester with the ball as the element receiving the impact.

5. The use of harvester with magnetic processing for simultaneous transmission of power and information through the supra-acoustic wave

The use of smart materials for wireless power transmission (and information) proved to be practical, and the results obtained during the research indicated the high efficiency of this method. Following further work, the project of Smart Ultrasonic Resonant Power System (SURPS) was created, which provides for the possibility of such transmission via various media and through various transmitter-receiver configurations. Diagrams are shown in **Figure 14**.

The mechanism of energy transmission consists in "sending" mechanical energy through the actuator in the form of a pure, sinusoidal ultrasonic wave and then its "pickup" by the harvester through the magneto- or electrostatic material that is in it. In this way, energy (along with information) can be transmitted not only through different types of centers but also at different distances. The type of frequency modulation (FM) was used for transmission of information, which for the needs of various types of structures was modified so that the data transfer was less than the resonant frequency of the structure. This method worked well during laboratory tests, and a flow chart is shown in **Figure 14**. **Figure 15A** shows a schematic diagram of data sent by an actuator based on Terfenol-D (AT) or piezoelectric material (AP). **Figure 15B** in turn shows the signal that is obtained on a harvester with a core of magnetostrictive material. **Figure 15C** illustrates the result of the operation of the station with two magnetostriction rails and transducers from **Figure 16**. A sinusoidal carrier frequency with small harmonic distortion (in the below 23 kHz case) generated by an actuator for data transmission to a harvester-powered microprocessor is modulated in the "on-off" mode, that is, in some time fragments, the actuator does not work by temporarily disconnecting the power supply of the harvesting side. Due to the fact that the harvester power supply has been equipped with a bank of capacitors with a capacity of 0.5 s microprocessor operation without harvest rami support, satisfactory results have been achieved even when transmitting many bytes of information encoded in accordance with ASCII signs.

Simultaneous supply of the sensory system was obtained, based on an industrial 32-bit microprocessor system and data transmission in half-duplex mode at the speed of about 1000 bps with the recovery of energy from mechanical vibrations with an over-acoustic frequency. A technique for feeding the microprocessor system from harvester machines combined with various configurations at carrier frequencies depending on the natural frequency of the structure containing dedicated actuators and harvester units with electromagnetic and magnetostriction transducers was developed.

Figure 16 shows a view of the assembled rail system with magnetostriction transducers with the possibility of powering the microprocessor system on the harvester side and transferring data in both directions.

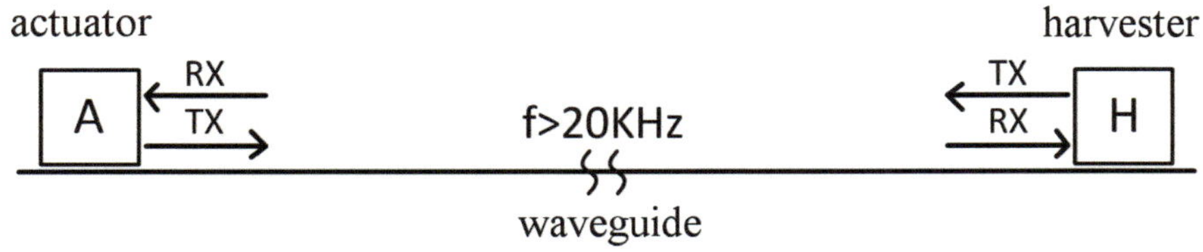

Figure 14.
A schematic diagram of power transmission through ultrasonic vibrations.

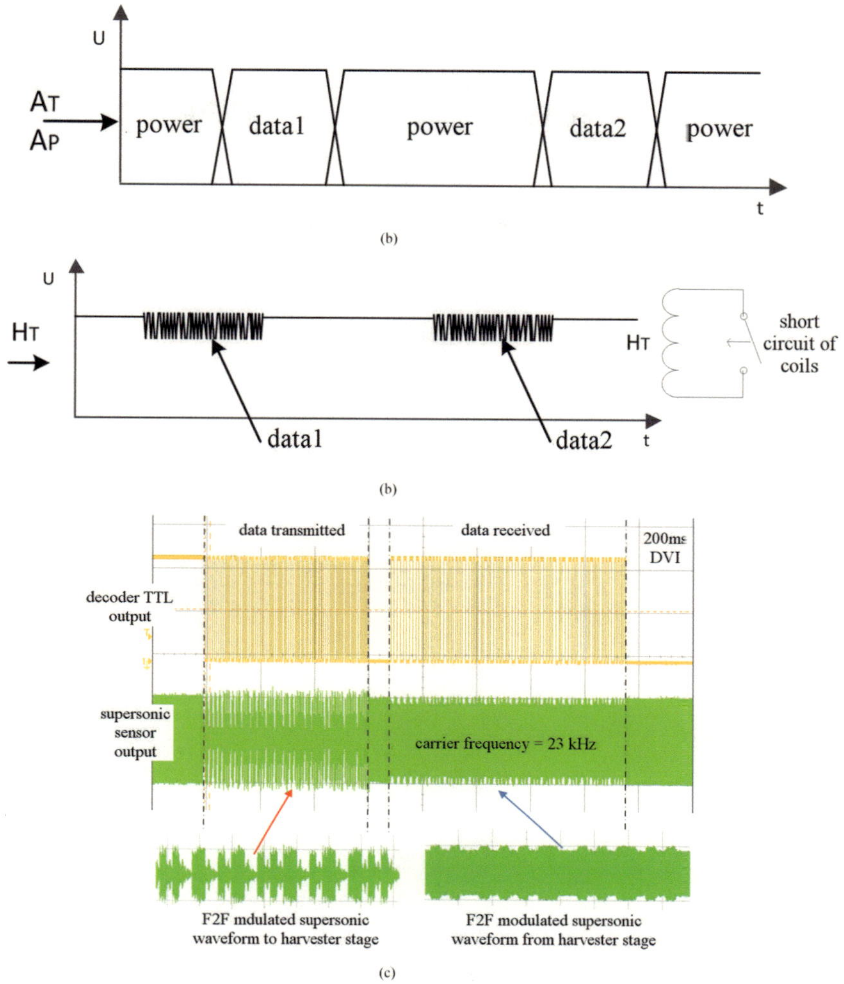

Figure 15.
Diagram of data sent (A) and received (B) by elements of the SURPS system. AT/P, actuator based on Terfenol-D/piezo material; HT, harvester based on Terfenol-D.

One of the main assumptions of the two rails' system was to set the required prestress, within a single structure, obtained by means of plastic elements, separately with each magnetostrictive transducer. This resulted in mutual mechanical coupling of the actuator and harvester and the possibility of adjusting the resonance frequency lying in the over-acoustic band.

Figure 17 shows the difference in the structure of the actuators based on magnetostriction and piezoelectric transducers. The characteristic differences relate to the way of generating the signal that powers the given actuator. In the case of magnetostrictive devices, in which the induction coil is loaded, the basic problem is to obtain a sufficient level of magnetostriction at a current that does not overheat the magnetic circuit with the core. Piezoelectric actuators require a voltage of 200VRMS, which is obtained through a bandwidth transformer with a primary winding matched to the power level based on the M-type metal-oxide semiconductor field-effect transistor (MOSFET) configuration in the H-configuration. During the development of the SURPS system, a structure of certain stages of electromagnetostrictive actuators and harvester was developed.

Based on the assumptions described above, as well as the current state of knowledge in the field of ultrasonic, wireless power transmission, a complete transceiver

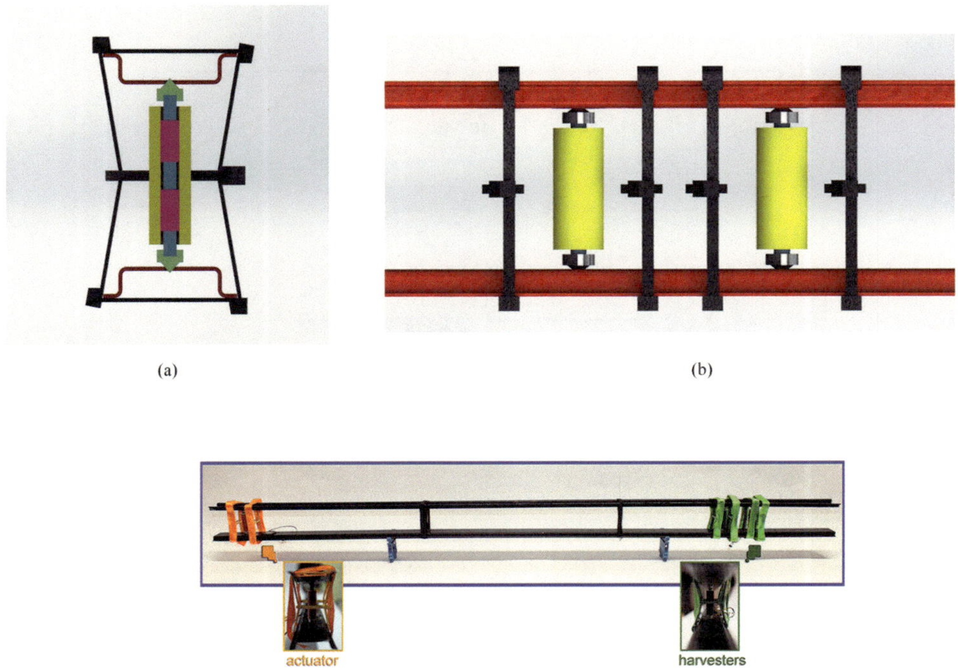

Figure 16.
View of the system of two rails with marked actuator and a unit recovering energy from mechanical vibrations (a) model, (b) real construction.

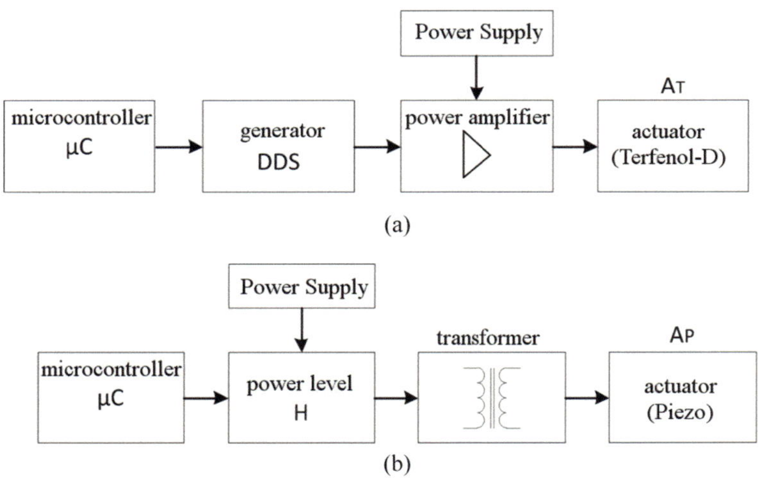

Figure 17.
Specification of the individual sections of the actuator: magnetostrictive (A), electrostatic element (B).

system was designed based on a suitable microcontroller, attendance modulators, as well as dedicated software.

The main features of the SURPS system are:

- Operation of piezoelectric actuators/harvester and magnetic processing.
- Finding and generating the resonance frequency of mechanical construction.

- Scanning of a given frequency range using the actuator-harvester system with real-time performance readout.
- Reading the current root mean square (RMS) voltage from the harvester.
- The system is equipped with the possibility of generating signals for two actuators generating vibrations of the same frequency but shifted in phase with each other.
- Data transmission between the actuator and return harvester sections (Tx, Rx).
- Frequency range from 0.1 to 50,000 Hz, 0.1 Hz (used direct digital synthesis generator (DDS) Analog Devices AD9851).

Figure 18 shows the frequency response of the mechanical structure of **Figure 16**. It is noteworthy that the highest performance (highest voltage) is in the over-acoustic range (above 20 kHz). The "SW" zone means an acceptable range of resonance frequencies lying near 20 kHz. On the characteristics with a dashed line, the 2.5 V voltage value is marked to guarantee the start of the microprocessor system. The points "A" and "B" marked on the waveform correspond to the most favorable ranges of carrier frequencies; it means that there are more frequencies capable of powering the system, and depending on the needs, the desired ranges of carriers can be selected. It is also possible to work more microprocessors connected to the same harvester but activated by a strictly defined frequency. The latter option allows the described solution to be used in SHM applications.

As a model microprocessor system, a Silicon Laboratory solution called Gecko with a 32-bit Cortex-M3 processor with the designation EFM32TG840 was used. In all applications, this type of set was used, and the solution had to guarantee the ability to supply this system as typically industrial with simultaneous half-duplex transmission (data transmission in one direction at a time in a bidirectional channel).

Data transfer is carried out using our algorithmy which we called frequency double frequency amplitude modulation (F2F-AM). As a result, the flow of information is much lower than the resonant frequency caused by ultrasounds or the structure itself and can get up to 1000 bps. Higher information flow rates can be obtained by using other types of frequency modulation.

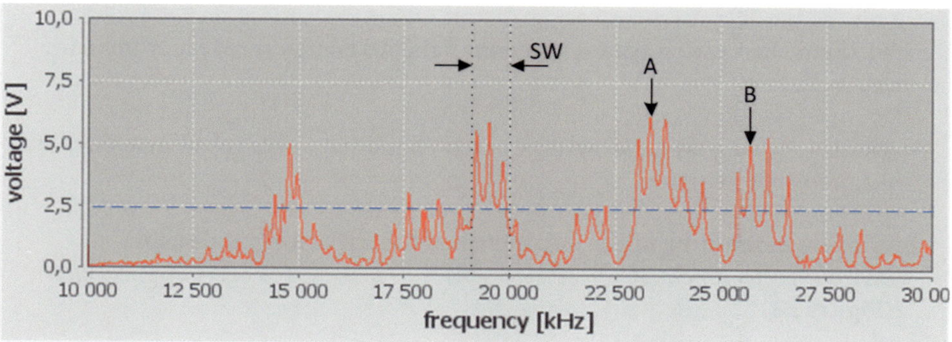

Figure 18.
The frequency response of the double bus system.

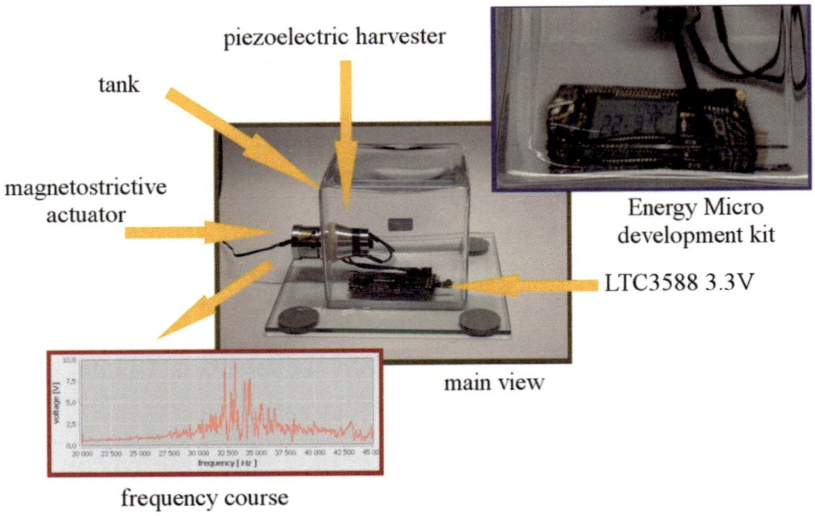

Figure 19.
View and frequency response of the system for simultaneous power transmission through the wall of a hermetic glass container with a wall thickness of about 10 mm.

Thanks to the observations made, it was found that the use of solutions based on the SURPS system developed under the author's guidance enables the transmission of energy over long distances without using cables but only through inaudible mechanical vibrations. However, the location of the harvester in different types of construction cannot be arbitrary. It is closely related to the medium in which the transmission takes place, as well as the length of the ultrasonic carrier wave.

A prototype of a simultaneous supply and data transmission system to the microprocessor sensory system was also created by the hermetic tank wall as a result of ultrasound wave stimulation, which is shown in **Figure 19**. Although its structure is similar, the frequency responses differ due to the different resonant frequency of the piezoelectric harvester. In this case, the actuator was a broadband magnetostrictive actuator. Behind the harvester, a piezoelectric cone transducer with a natural frequency of 38 kHz is placed.

6. Autonomous system of diagnostics of environmental and operating parameters named Multi-DOF

Multi-node harvesting systems for simultaneous energy recovery from many sources, including:

- Multi-node harvesting structures based on miniature harvester machines with magnetostrictive cores
- Wireless monitoring of the parameters of the harvesting node using micro-electro-mechanical systems' (MEMS) sensors for SHM applications (**Figure 20**)
- Microprocessor systems powered from harvesting sources
- Autonomous monitoring system Multi-DOF

Figure 20.
View of a single electronic wireless node powered from a harvesting source.

In the field of low-power technology, the definition of harvester as a single microprocessor power supply (μC) was adopted, which after wireless feeding sends data in accordance with its operating algorithm (program code) to the receiving and processing unit. A single harvesting system is a node in a larger structure managed from a central site. Individual configurations of harvester can allow tuning the harvesting power supply to specific phenomena that trigger its operation. **Figure 21** shows a schematic diagram of a harvesting structure consisting of several harvesters activated as a result of an external stimulus.

Harvesters which in their principle of work use cross effects, more and more often are based on magneto-mechanical phenomena. It is assumed that even in the case of low power and efficiency, they can be a valuable source of power supply.

Multi-node harvesting structure can be used in structural health monitoring (SHM) applications to recover an electric power from the wasted energy generated mostly from vibrations. Magnetic harvester also might be used as a power source in SHM systems which are monitoring large mechanical structures. Our latest system presents this solution. It uses 14 MEMS sensors which designated 14 degrees of freedom (DOF) (3D accelerometer, 3D gyroscope, 3D magnetometer, barometric pressure sensor, microphone, temperature T, humidity R, light intensity). The structure of the system was shown in the **Figure 22**. The software designed by authors allows to monitor the parameters provided by 14 sensors via web page or

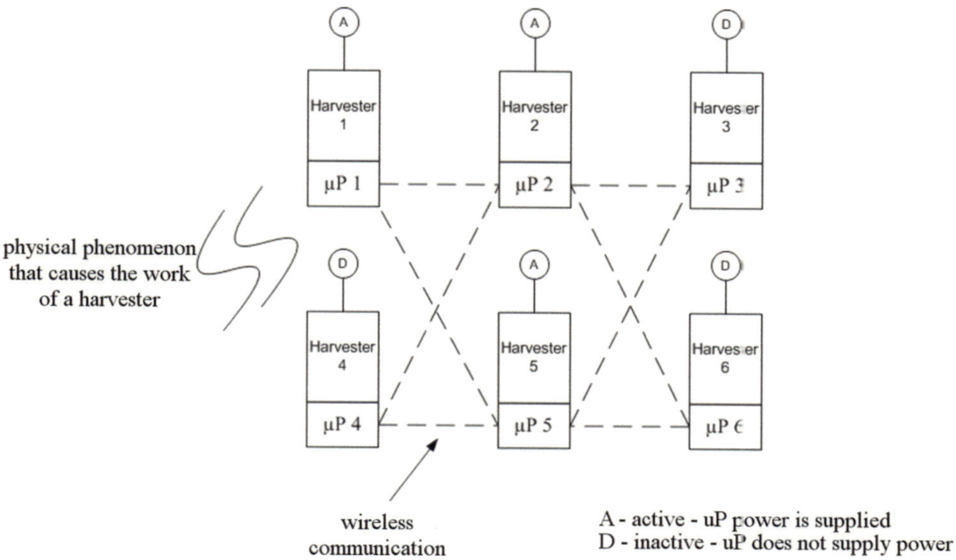

Figure 21.
Activation of harvester by physical phenomenon.

in service mode. The software is designed to support such systems as ADIS16488 module and other components of one of the most precise Analog Devices iMEMS 2016 (IMU). In order to process data received from the 14DOF sensors, which include not only measuring the certain physical value but also monitoring the level of recovered energy, the proper microprocessors had to be chosen (an important factor is a power consumption).

Figure 23 shows three typical sources of low-frequency energy harvesting: mechanical shock wave (**Figure 23A**), low-frequency mechanical resonance (**Figure 23B**), and energy transmission though ultrasonic resonant vibrations (**Figure 23C**). Properly selected conditioning circuit provides the harvesting system with a useful current and voltage capabilities. The creation of a wireless node to measure certain physical quantities and to monitor the level of recovered energy

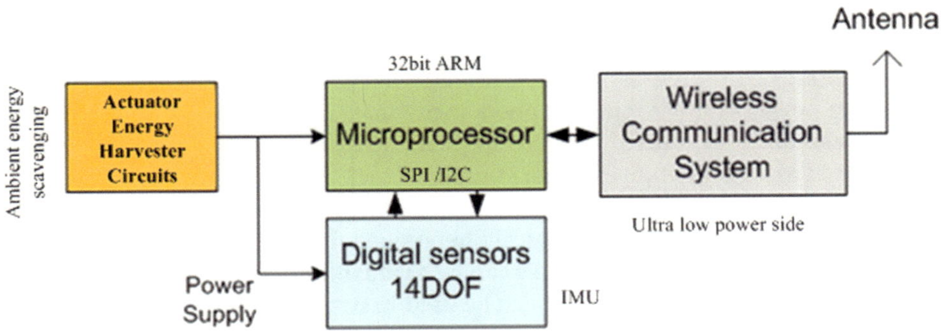

Figure 22.
The structure of a wireless harvesting system with a 14DOF block.

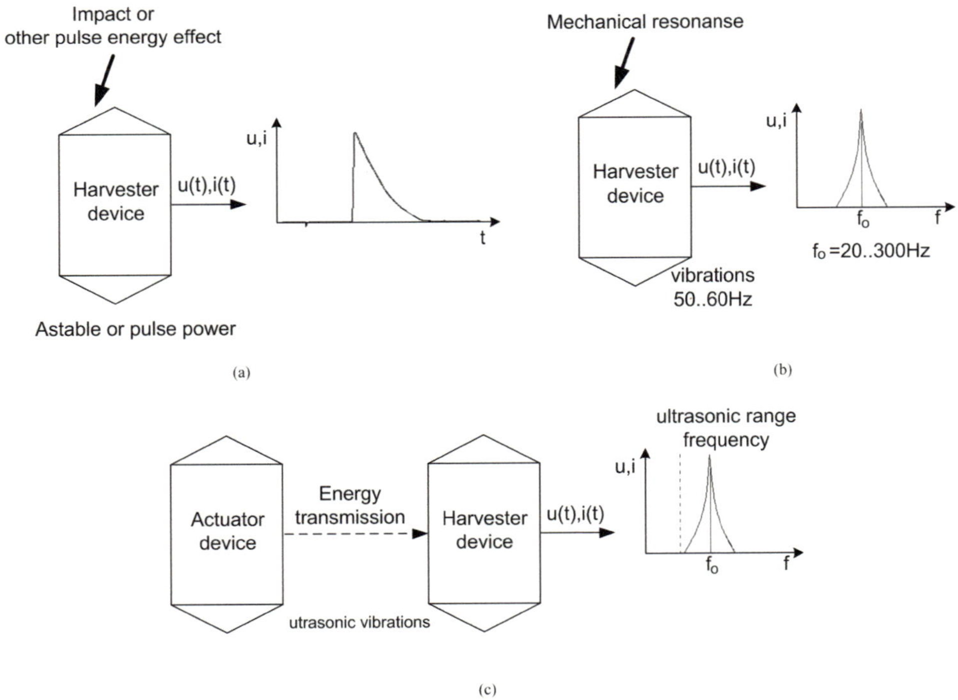

Figure 23.
Energy-harvesting sources and their power requirements: (A) mechanical impact, (B) low-frequency mechanical resonance, and (C) energy transmission by ultrasonic vibration.

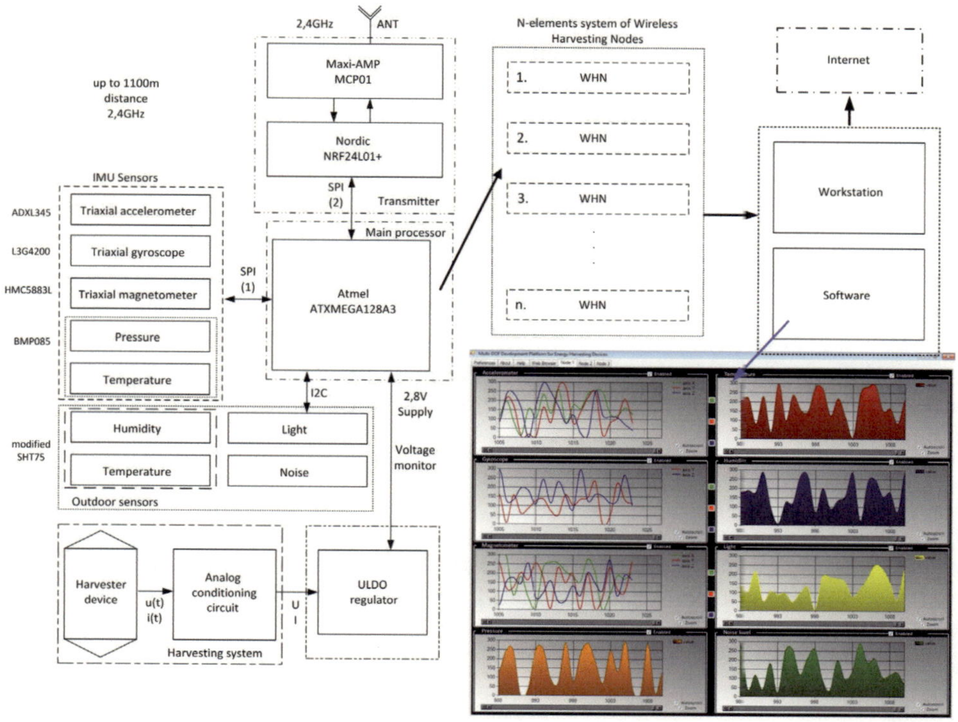

Figure 24.
Prototyping of the multi-DOF wireless sensor platform: main communication station and the Multi-DOF software.

requires selection of an appropriate hardware platform such as a microprocessor and wireless transmission system. The use of smart materials in wireless power transmission turned out to be effective. For this purpose, a SURPS system for simultaneous power and data transmission was developed. It ensured transmission through various media (solid, liquid) and with various transmitter-receiver configurations [X].

After matching the sensor-microprocessor configuration with a suitable energy harvester, the whole packets, together with a wireless communication system, were placed in the nodes. Due to the fact that every node is equipped with the same wireless communication system, different types of sensors can be easily substituted or put together by the user, thanks to the dedicated software shown in **Figure 24**.

Properly selected conditioning circuit provides the harvesting system with a certain current and voltage output. The creation of a wireless node to measure certain physical quantities and to monitor the level of recovered energy requires selection of an appropriate hardware platform such as a microprocessor and wireless transmission system.

7. Conclusions and final remarks

The essence of EH is to create new concepts of current generators, using cross effects, including more often magnetomechanical phenomena. The use of smart materials for wireless power transmission (and information) proved to be practical, and the results obtained during the research indicated the high efficiency of this method.

A technique for powering the microprocessor system from harvester machines combined with various configurations at carrier frequencies depending on the natural frequency of the structure containing dedicated actuators and harvester units with electromagnetic and magnetostriction transducers was developed. Satisfactory results have been achieved even when transmitting many bytes of information encoded in accordance with ASCII characters. Simultaneous supply of the sensory system was obtained, based on an industrial 32-bit microprocessor system and data transmission in half-duplex mode at the speed of about 1000 bps with the recovery of energy from mechanical vibrations with an over-acoustic frequency.

Thanks to the observations made, it was found that the use of solutions based on the SURPS system developed by the authors enables the transmission of energy over long distances without using cables but only through inaudible mechanical vibrations. However, the location of the harvester in different types of construction cannot be arbitrary. It is closely related to the medium in which the transmission takes place, as well as the length of the ultrasonic carrier wave.

Acknowledgements

The research was funded by the National Center for Research and Development within LIDER IX project (grant number: LIDER/21/0082/L-9/17/NCBR/2018).

Author details

Jerzy Kaleta*, Rafał Mech and Przemysław Wiewiórski
Department of Mechanics, Materials Science and Engineering, Wrocław University of Science and Technology, Wrocław, Poland

*Address all correspondence to: jerzy.kaleta@pwr.edu.pl

References

[1] Roundy S, Wright PK, Rabaey JM. Energy Scavenging for Wireless Sensor Networks: With Special Focus on Vibrations. New York: Springer; 2004

[2] Roundy S, Steingart D, Frechette L, Wright P, Rabaey J. Power sources for wireless sensor networks. In: Karl H, Wolisz A, Willig A, editors. Lecture Notes in Computer Science, nr 2920. Springer Berlin Heidelberg; 2004. pp. 1-17

[3] Mitcheson PD, Yeatman EM, Kondala Rao G, Holmes AS, Green TC. Energy harvesting from human and machine motion for wireless electronic devices. In: Proceedings of the IEEE. Vol. 96. 2008. pp. 1457-1486

[4] Kompis C, Aliwell S. Energy Harvesting Technologies to Enable Remote and Wireless Sensing; 2008

[5] Green PL. Nonlinear Energy Harvesting, Rozprawa Doktorska. Department of Mechanical Engineering, University of Sheffield; 2012

[6] Energy Autonomous Systems: Future Trends in Devices, Technology, and Systems. 2019. Available from: http://www.catrene.org/web/about/EAS_ExecutiveSumm_v7CAT.pdf

[7] Murry EJ. A unique system for transmission of ultrasonic energy over fibrous bundles. Ultrasonics. 1970;**8**(3):168-173

[8] Sherman ML, Castellano TM. Ultrasound energy delivery system and method, US patent 5,735,280; 1989

[9] Denisov A, Yeatman E. Stepwise microactuators powered by ultrasonic transfer. Procedia Engineering. 2011;**25**:685-688

[10] Ozeri S, Shmilovitz D. Ultrasonic transcutaneous energy transfer for powering implanted devices. Ultrasonics. 2010;**50**(6):556-566

[11] Mazzilli F, Thoppay PE, Praplan V, Dehollain C. Ultrasound energy harvesting system for deep implanted-medical-devices (IMDs). In: 2012 IEEE International Symposium on Circuits and Systems (ISCAS); IEEE. 2012. pp. 2865-2868

[12] Ottman GK, Hofmann HF, Bhatt AC, Lesieutre GA. Adaptive piezoelectric energy harvesting circuit for wireless remote power supply. In: IEEE Transactions on Power Electronics. 2002. pp. 669-676

[13] Ishiyama T, Kanai Y, Ohwaki J, Mino M. Impact of a wireless power transmission system using an ultrasonic air transducer for low-power mobile applications. In: 2003 IEEE Symposium on Ultrasonics; IEEE. 2003. pp. 1368-1371

[14] Li P, Wen Y. Energy harvesting transducer by collecting electromagnetic energy based on ultrasonic horn. In: IEEE International Conference on Information Acquisition; IEEE. 2006. pp. 550-555

[15] Roes MGL, Hendrix MAM, Duarte JL. Contactless energy transfer through air by means of ultrasound. In: IECON 2011—37th Annual Conference on IEEE Industrial Electronics Society; IEEE. 2011. pp. 1238-1243

[16] Lomonova E, Roes M, Duarte J, Hendrix M. Acoustic Energy Transfer: A Review; 2013

[17] Kural A, Pullin R, Featherston CA, Lees J, Naylon J, Paget C, et al. Wireless power transmission using ultrasonic guided waves-electric circuit measurement and simulation. Key Engineering Materials. 2012;**518**:445-454

[18] Kural A, Pullin R, Featherston C, Lees J, Naylon J, Paget C, et al. Wireless electric power transmission using ultrasonic guided waves. In: Energy Harvesting. 2012. Available from: http://eh-network.org/events/eh2012.php

[19] Saulnier GJ, Scarton HA, Gavens AJ, Shoudy DA, Murphy TL, Wetzel M, et al. P1g-4 through-wall communication of low-rate digital data using ultrasound. In: 2006 IEEE Ultrasonics Symposium; IEEE. 2006. pp. 1385-1389

[20] Lawry T. A High Performance System for Wireless Transmission of Power and Data through Solid Metal Enclosures, Rozprawa Doktorska; Rensselaer Polytechnic Institute; 2011

[21] Lawry TJ, Wilt KR, Ashdown JD, Scarton HA, Saulnier GJ. A high-performance ultrasonic system for the simultaneous transmission of data and power through solid metal barriers. In: IEEE Transactions on Ultrasonics, Ferroelectrics and Frequency Control. 2013. pp. 194-203

[22] Lawry TJ, Wilt KR, Scarton HA, Saulnier GJ. Analytical modeling of a sandwiched plate piezoelectric transformer-based acoustic-electric transmission channel. In: IEEE Transactions on Ultrasonics, Ferroelectrics and Frequency Control. Vol. 59. 2012. pp. 2476-2486

[23] Sherrit S, Mukherjee BK. The use of complex material constants to model the dynamic response of piezoelectric materials. In: Proceedings Ultrasonics Symposium; IEEE. 1998. pp. 633-640

[24] Sherrit S, Badescu M, Bao X, Bar-Cohen Y, Chang Z. Efficient electromechanical network model for wireless acoustic-electric feed-throughs. In: Smart Structures and Materials. International Society for Optics and Photonics; 2005. pp. 362-372

[25] Sherrit S, Doty B, Badescu M, Bao X, Bar-Cohen Y, Aldrich J, et al. Studies of acoustic-electric feed-throughs for power transmission through structures. In: Smart Structures and Materials; International Society for Optics and Photonics. 2006. pp. 617102

[26] Sherrit S. The physical acoustics of energy harvesting. In: 2008 IEEE Ultrasonics Symposium; IEEE. 2008. pp. 1046-1055

[27] Sherrit S, Walkemeyer P, Bao X, Bar-Cohen Y, Badescu M. Acoustic mechanical feedthroughs. In: SPIE Smart Structures and Materials+ Nondestructive Evaluation and Health Monitoring; International Society for Optics and Photonics. 2013. p. 86920P

[28] Mirosławski M. Stan prac z dziedziny Energy Harvesting w Laboratorium Dynamiki; stan wiedzy z dziedziny bezprzewodowego, ultradźwiękowego przesyłu energii na świecie, Rap. Tech., Instytut Materiałoznawstwa i Mechaniki Technicznej Politechniki Wrocławskiej, Wrocław; 2013

[29] Beeby SP, Tudor MJ, White NM. Energy harvesting vibration sources for microsystems applications. Measurement Science and Technology. 2006;**17**(12):R175

[30] Mitcheson PD, Green TC, Yeatman EM, Holmes AS. Architectures for vibration-driven micropower generators. Journal of Microelectromechanical Systems. 2004;**13**(3):429-440

[31] Beeby S, White NM. Energy Harvesting for Autonomous Systems. London: Artech House; 2010

[32] Torah R, Glynne-Jones P, Tudor M, O'Donnell T, Roy S, Beeby S. Self-powered autonomous wireless sensor node using vibration energy harvesting.

Measurement Science and Technology. 2008;**19**(12):125202

[33] Kaleta J, Kot K, Mech R, Wiewiórski P. Ultrasonic energy harvesting system based on magneto- and electrostrictive actuators. In: The 7th International Symposium on Mechanics of Materials and Structures, Conference Proceedings; Augustów, Poland. 2013. pp. 133-137

[34] Ghemari A. The usage of selected methods of Energy Harvesting in mechanics; construction and applications [Master thesis]. Wrocław: Politechnika Wrocławska, Wydział Mechaniczny; 2010

[35] Kaleta J, Lewandowski D, Wiewiórski P, Mech R, Liberda M. Power generating by high pulse mechanical stimulation of magnetic coupled NdFeB and Terfenol-D. In: SPIE Smart Structures and Materials + Nondestructive Evaluation and Health Monitoring; International Society for Optics and Photonics. 2010. p. 76440U

[36] Kaleta J, Lewandowski D, Liberda M, Mech RP, Wiewiórski PK. Energy harvester with high pulse mechanical stimulation. In: 27th Danubia-Adria Symposium on Advances in Experimental Mechanics. Wrocław: Wrocław University of Technology; 2010

[37] Kaleta J, Kot K, Rikitatt M, Wiewiórski P. Multid. of wireless sensor system based on iMU-MEMS technology supported by energy harvesting methods. In: 4th International Conference on Integrity, Reliability & Failure; Funchal. 2013

[38] Bomba J. Investigation of GMM actuator as a communications system. In: Communications System Application of GMM. 2006. na prawach rękopisu

[39] Bomba J, Kaleta J. Sprawność przekształcania energii w materiałach o gigantycznej magnetostrykcji. In: IX Krajowa Konferencja Mechaniki Pękania. 2003. pp. 47-54

[40] Lewandowski D. Solid State Magnetic Phenomena Harvesters and their Power Conditioners for Low Power Applications. Berlin, Germany: Harvesting & Storage Europe; 2013

[41] Kaleta J, Wiewiórski P. Rapid demagnetization of neodymium magnets, due to mechanical shock, as a pulse power supply for microprocessor system. In: International Conference on Integrity, Reliability & Failure; Funchal. 2013

[42] Energy Harvesting: Downloads. 2019. Available from: http://www.energyharvesting.pl/download.html

[43] Holt TA. Explosively-Driven Helical Magneto-Cumulative Generators, Praca Magisterska. Texas Tech University; 2002

[44] Lee J, Choi JS, Ryu JH, Kim CH. Maximizing the energy output of explosively-driven ferromagnetic generators. Electromagnetic Phenomena. 2003;**3**(3):11

Chapter 4

Small-Scale Energy Harvesting from Environment by Triboelectric Nanogenerators

Jie Wang, Linglin Zhou, Chunlei Zhang and Zhong Lin Wang

Abstract

The increasing needs to power trillions of sensors and devices for the Internet of Things require effective technology to harvest small-scale energy from renewable natural resources. As a new energy technology, triboelectric nanogenerators (TENGs) can harvest ambient mechanical energy and convert it into electricity for powering small electronic devices continuously. In this chapter, the fundamental working mechanism and fundamental modes of a TENG will be presented. It can harvest all kinds of mechanical energy, especially at low frequencies, such as human motion, walking, vibration, mechanical triggering, rotating tire, wind, moving automobile, flowing water, rain drops, ocean waves, and so on. Such variety of energy harvesting methods promises TENG as a new approach for small-scale energy harvesting.

Keywords: mechanical energy harvesting, triboelectric nanogenerators, biomechanical energy, vibration energy, wind energy, water energy, self-powered system

1. Introduction

With the rapid development of science and technology, Internet of Things (IoT) plays an increasingly important role in the next evolution of the Internet through turning data into information, knowledge, and wisdom [1]. More recently, multiple type applications based on IoT have been developed, including health testing, safe home, intelligent transportation, logistics supply, environmental protection, infrastructure testing, and security [2]. Sensor nodes in the IoTs are widely distributed and require independent, mobile, sustainable, and maintenance-free capabilities. Under the current technologies, most sensors require an external power source to drive their operation, wherein the battery is extensively applied. However, the life cycle of the battery is limited, and replacing the battery for the massive sensors is a huge project, which consumes a lot of manpower and material resources and increases the maintenance cost. In addition, the regularly replaced battery generates a large amount of harmful substances, which seriously endangers the environment and human health. Therefore, a clean and sustainable power source should be provided to satisfy the requirement of driving these small electronic devices sustainably.

Harvesting of the ambient environment energy, as an eco-friendly and renewable collecting energy method, is regarded as a promising and effective strategy to realize continuous powering for these small electronic equipment [3]. Some possible technologies have been exploited for collecting energy from surrounding environment, such as solar cells that collect energy from sunlight [4] and thermoelectric generators that harvest energy from temperature difference [5]. However, as constrained by the intermittency nature of sunlight, the low output of thermoelectric generators, these energy harvesting technologies cannot ensure the continuous operation of electronic devices. Owing to its abundant reserves and widespread, mechanical energy are increasingly utilized to extract and convert into electricity based on different mechanisms, including electromagnetic generator (EMG) [6], piezoelectric nanogenerator (PENG) [7, 8], and triboelectric nanogenerator (TENG) [9]. Considering the large-scale power generation of EMG and low output power of PENG, TENG has been demonstrated as a promising approach for harvesting ambient mechanical energy due to the desirable features of simple structure, flexibility, low cost, light weight, high efficiency, and high power density at low frequency [10]. The operation of TENGs is depended on triboelectrification (or contact electrification) and electrostatic induction [11], and the fundamental theory is according to Maxwell's displacement current and change in surface polarization [12]. Since the first invention of TENG in 2012, it has been extensively investigated and well confirmed that the potential of wide application is ranging from powering small electronic devices for self-powered systems, functioning as active sensors for medical, infrastructural, human-machine, environmental monitoring, and security [13–20]. Various types of wasted mechanical energies in our daily life, such as human motion, vibration, wind, and flowing water can be utilized by different TENG structures. Based on these characteristics, TENG can be utilized as a small-scale energy harvester for driving mass electronic equipment continuously.

2. Fundamental working modes of the TENG

TENGs are derived from the coupling effect of contact electrification and electrostatic induction. Contact electrification, as known as static electricity and contact charging, is a common phenomenon in many manufacturing environments and has been known for thousands of years. During the process of contact electrification, the dissimilar material/surface becomes charged after contacting with each other. After contacting, the opposite's triboelectric charge is produced on the surface of dissimilar materials with different electron affinities. Driven by external mechanical motion, the materials will be separated resulting in potential difference between the two electrodes on the back side of the materials. To maintain the electrostatic equilibrium, the free electrons in the electrodes will be driven to flow in external circuit to balance the induced potential difference, thus converting mechanical energy into electrical energy. According to the different structure designs of electrodes or moving manners of the triboelectric layer in TENGs, four different modes of TENGs have been build [9], as elaborated as follows.

2.1 Vertical contact-separation mode

The mechanism of vertical contact-separation mode can be elaborated largely by an example. As shown in **Figure 1a**, the simplest structure of TENG includes

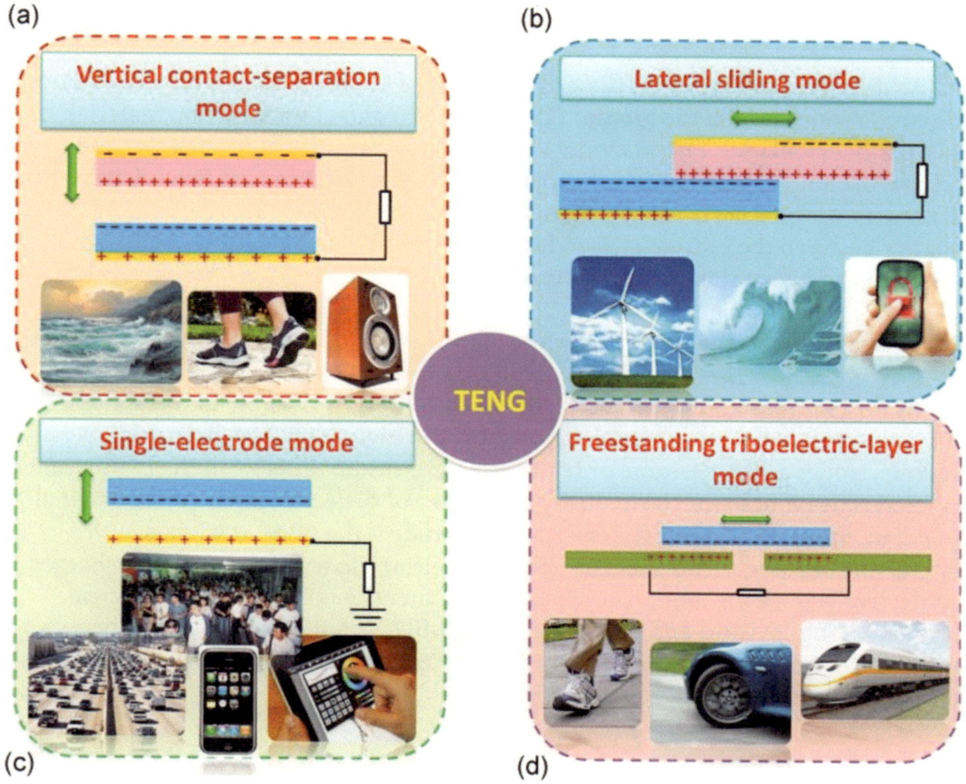

Figure 1.
The four fundamental modes of triboelectric nanogenerators: (a) vertical contact separation mode, (b) in-plane contact-sliding mode, (c) single-electrode mode, and (d) freestanding triboelectric-layer mode [9].

two metal electrodes and a dielectric film, in which two metal films work as top electrode and back electrode attached to dielectric film, respectively [21, 22]. When mechanical movement is applied in the unit, the top electrode and dielectric film will contact with each other, and thus the dielectric layer and electrode will get positively charged and negatively charged, respectively, due to the triboelectrification. Once they are separated by a short distance, the potential difference between the two electrodes will be induced, which will drive electrons to flow from the back electrode to the top electrode, resulting in a pulse current with an external circuit connected. If they are brought into contact again, the electrons will flow back and the current will be reversed.

2.2 Lateral-sliding mode

The basic structure of TENG in this model is the same as that of the vertical contact-separation mode. The difference is from the motion mode of the top electrode (**Figure 1b**). In the original state, the top electrode and dielectric film fully overlap and intimately contact with each other, leading to the oppositely charged surfaces. With the top electrode sliding outward, the contact surface area will decrease gradually until the complete separation of the two surfaces. The separated surface creates a potential difference across the two electrodes, generating a current flow from the top electrode to the bottom electrode. When it slides backward, then there will be a reversed current flow to balance the potential difference [23, 24].

2.3 Single-electrode mode

As displayed in **Figure 1c**, the single-electrode mode TENG has only one bottom electrode connected to the ground [25, 26]. After contact with the top material, the two surfaces will get charged owing to the triboelectric effect. During the process of an approaching and departing of top material, the local electrical field distribution caused by charged surfaces will change. Then, there will be potential difference change between the bottom electrode and the ground, and electrons exchange between them to maintain the potential change.

2.4 Freestanding triboelectric-layer mode

As for the freestanding triboelectric-layer mode, it is the only one that the motion part is a dielectric layer [10], as shown in **Figure 1d**. The dielectric layer and two electrodes are in the same order, and the gap distance between the two symmetric electrodes should much smaller than the size of dielectric layer. At the original position, the state of dielectric layer and electrode is the same as what is in the lateral-sliding mode. The dielectric layer and electrode will get oppositely charged, respectively, once the motion occurs as before mentioned. When the dielectric layer is sliding forward and backward, there will be a potential difference between the two electrodes due to the change of overlapped area, which drives the electron exchanges between them.

3. TENGs as small-scale power source

In order to satisfy the requirement of harvesting mechanical energy from multiple type motions, various TENGs have been fabricated based on the four modes illustrated above.

3.1 TENGs harvesting energy from biomechanical

Given the collection features of small scale, low frequency, and irregularity, human biomechanical motions are considered to be accessible, renewable, and the most abundant energy sources. TENG can collect this energy and convert it into electricity. Since it is first reported in 2012, TENG harvesting mechanical energy from human biomechanical movements has been fully developed.

Compared to the discrete devices, complex integrated TENGs can perform multiple functions with the merits of higher output performance, better adaptability, and sustainably. Based on the high-efficient and sustainable TENGs, various integrated TENGs have been developed for harvesting energy from human biomechanical movements. Zhu et al. introdued a packaged power-generating insole with built-in flexible multi-layered TENGs that harvested mechanical pressure during normal walking to power portable and wearable consumer electronics [27]. Bai et al. developed a flexible multilayered TENG by intergrating five layers of units on a zigzag-shaped Kapton substrate to gain pressure from normal walking [28]. Because of the unique structure and nanopore-based surface modification on the metal surface, the instantaneous short-circuit current (Isc) and the open-circuit voltage (Voc) can reach 0.66 mA and 215 V with an instantaneous maximum power density of 9.8 mW/cm^2 and 10.24 mW/cm^3. Triggered by press from normal walking, the TENG attached onto a shoe pad was able to instantaneously drive multiple commercial LED bulbs.

For improving the output current, Yang et al. designed an integrated rhombic gridding-based TENG to harvest vibration energy from natural human walking [29]. The newly designed TENG consists of PTFE nanowire arrays and aluminum nanopores with the hybridization of both the contact-separation mode and sliding electrification mode. Herein, Voc of the TENG could be up to 428 V, and Isc was near 1.395 mA with the peak power density of 30.7 W/m^2. Moreover, based on the TENG, a self-powered backpack was developed with a considerably high vibration-to-electric energy conversion efficiency of 10.62(±1.19)%. When a person walks naturally carrying the designed backpack with a total weight of 2.0 kg, the power harvested from the body vibration is high enough to simultaneously light all the 40 LEDs.

Based on a high-output TENG, Niu et al. developed an universal self-charging system exclusively driven by random body motion for sustainable operation of mobile electronics [14]. In this system, a multilayered attached-electrode contact-mode TENG is utilized to effectively collect the energy from human walking and running (**Figure 2a**). The basic working principle of attached-electrode contact-mode TENGs is shown in **Figure 2b**. The structure of multi-unit TEMG, shown in **Figure 2c**, consists of 10–15 layered TENGs which used a Kapton film (a thickness of 125 μm) as the substrate and is shaped into a zigzag structure. A surface modified thin aluminum foil and fluorinated ethylene propylene (FEP) layer are

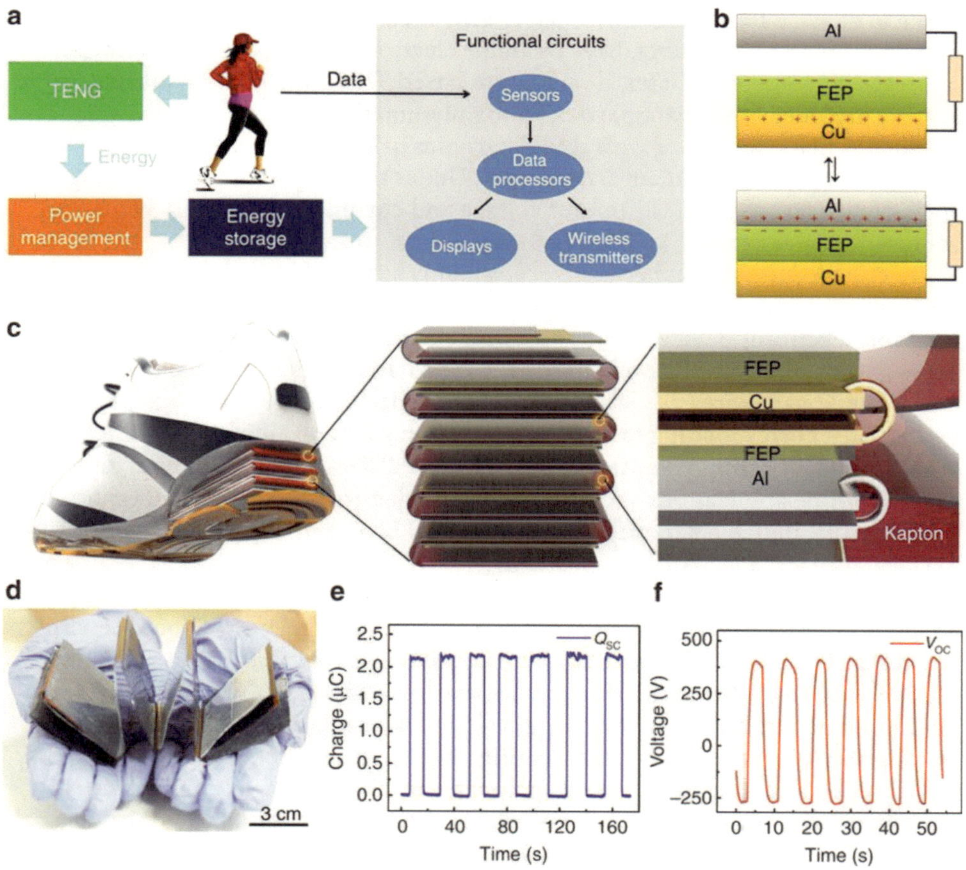

Figure 2.
(a) System diagram of a TENG-based self-powered system, (b) working mechanism of an attached-electrode contact-mode TENG, (c) structure of the designed multilayer TENG, (d) photo of an as-fabricated TENG, (e) triboelectric charge output, and (f) Voc output of the as-fabricated TENG [14].

utilized as the triboelectric materials. **Figure 1d** displays the small volume and lightweight of as-fabricated TENG (5.7 × 5.2 × 1.6 cm/29.9 g for a 10-layer TENG and 5.7 × 5.2 × 2.4 cm/43.6 g for a 15-layer TENG). As shown in **Figure 2e,f**, a human walking can drive this TENG to generate about 2.2 μC short-circuit transferred charge and about 700 V voltage output when embedded the TENG in the shoe insoles.

Shen et al. proposed a humidity resisting triboelectric nanogenerator to harvest energy from human biomechanical movements and activities for wearable electronics [30]. The obtained HR-TENG is fabricated by a nanofibrous membrane via electrospinning method. Under a relative humidity of 55%, the current and voltage output of the self-powered unit can still reach as high as 28 μA and 345 V, corresponding to a power density of 1.3 W/m^2 with hand tapping. With the relative humidity raising from 30 to 90%, its electrical output still kept a relatively high level. A wide-range of electronics such as an electronic watch, a commercial calculator, a thermal meter, and a total of 400 LEDs has demonstrated to be successfully powered from human biomechanical movements under different ambient humidities.

Textile-based device is highly desirable for wearable electronics due to its low-mass, durable, flexible, and conformable [31]. As the most efficient power sources, textile substrate-based TENGs are fabricated for the features of simple structure, wide material choices, and low cost [32–37]. Series efforts have been made to develop fabric TENGs for harvesting mechanical energy induced from body motions to sustainably drive wearable electronics [34, 38]. Lee et al. reported an electrical response of a textile substrate-based TENG including nanostructured surface provided by Al nanoparticles and polydimethylsiloxane (PDMS) [32]. The obtained TENG can power wearable electronics using low-frequency mechanical movements driven by human arm activity. Under the simple folding-releasing stage of an arm near 90°, the output voltage and current of 139 V and 39 μA are achieved, respectively.

To enhance the output performance, a highly stretchable 2D fabric was developed as a wearable TENG for harvesting footstep energy during walking to driven wearable electronic devices [39]. The fabric-structured TENG composes by Al wires and PDMS tubes with a high-aspect-ratio nanotextured surface with vertically aligned nanowires. It shows a stable high-output voltage and current of 40 V and 210 μA, corresponding to an instantaneous power output of 4 mW. The TENG also exhibits high robustness behavior even after 25% stretching, enough for use in smart clothing applications and other wearable electronics. Seung et al. reported a fully flexible, foldable nanopatterned wearable TENG with high power-generating performance and mechanical robustness [40]. Both a silver (Ag)-coated textile and PDMS nanopatterns based on ZnO nanorod arrays on a Ag-coated textile template are used as active triboelectric materials. A high voltage and current output with an average value of 170 V and 120 μA, respectively, are obtained from a four-layer-stacked wearable TNG under the compressive force of 10 kgf. Notably, there are no significant differences in the output voltages measured from the multilayer-stacked WTNG over 12,000 cycles, confirming the excellent mechanical durability of WTNGs. Without external power sources, the fabricated wearable TENG can drive the LEDs, LCD, and the keyless vehicle entry system, exihibting the potential applications in self-powered smart clothes, health care monitoring and self-powered wearable devices, and even personal electronics. Tian et al. demonstrated a high-performance double-layer-stacked triboelectric textile (DTET) for harvesting human motion energy [41]. Both the Ni-coated polyester conductive textile and the silicone rubber are adopted as effective triboelectric materials. A high output Voc of 540 V and an Isc of 140 μA can be obtained from the DTET with the size of

5 × 5 cm^2, corresponding to a high peak surface power density of 0.892 mW/cm^2 at a load resistance of 10 MΩ. The output peak signal of the DTET can be used as a trigger signal of a movement sensor to design movement monitoring equipment. With only the energy harvested from walking, running, or flapping, the DTET can directly light up 100 LEDs connected serially and drive portable electronics, such as competition timer, digital clock, and electronic calculator.

Owing to the high power density, stable cycle life, good safety, and potentials in integration into flexible wear, introducing supercapacitors as energy-storing devices into a fabric TENG show promising prospects. Pu et al. introduced a self-charging power textile for harvesting human motion energy. The self-charging power textile was fabricated by weaving the yarn supercapacitors together with a fabric TENG into an individual fabric [42]. Based on the integrated system, the motion-charging process is carried out by charging the yarn supercapacitors by the contact-separation motions between the TENG cloth and a common cotton cloth. The yarn supercapacitors and the fabric TENG endowed the excellent flexibility and weaveability of the self-charging power textile. Chen et al. developed a self-charging power textile, consisting of a fabric triboelectric nanogenerator and a woven supercapacitor, which can simultaneously harvest and store body motion energy to sustainably drive wearable electronics [43]. Utilizing traditional weaving craft, contact-separation mode and free-standing mode FTENG are designed and fabricated on a piece of textile by weaving the cotton, carbon, and PTFE wires. Combined with the energy-storing component, utilizing RuO_2-coated carbon fiber and cotton threads, the obtained self-charging power textile can harvest energy from common daily activities such as running and walking to drive the wearable electronics, such as an electric watch.

For developing low-cost TENG, paper served as a supporting component for preparing TENG for the first time [44]. Paper-based TENGs represent an low-cost, light-weight, and environmentally friendly energy harvesting methodology. Nowadays, different types of paper-based TNEG have been designed and prepared for harvesting energy from human biomechanical movements [45]. Xia et al. proposed a X-shaped paper TENG formed from a ballpoint ink layer coated by painting with a commercial brush pen for harvesting mechanical energy from human walking [46]. In this design, paper served as both a component of the triboelectric pairs and a supporting component. When a brush pen is painted on the paper, the maximum values of current and voltage output can be achieved at 326 V, 45 μA, corresponding to a power density of 542.22 μW/cm^2. The staked X-shaped paper TENG is proposed to increase the output performance and harvest the mechanical energy generated by motion of the human body, which can directly light up 101 blue high-power LEDs with a working voltage of 3.4 V.

Additionally, various efforts have been made to promote the development of TENGs for harvesting biomechanical energy based on external devices attached to the human body. In them, human skin-based TENGs are developed for converting biomechanical energy induced from human body itself into electronic energy. According to these series TENGs, human skin is used as one of the triboelectric materials with the single-electrode-mode. With the contact/separation between an area of human skin and a PDMS film, a Voc up to −1000 V, a short-circuit current density of 8 mA/m^2, and the corresponding power density of 500 mW/m^2 on a load of 100 MΩ were obtained from the skin-based TENG delivers, which could be used to directly drive tens of green light-emitting diodes [47]. Due to its fantastic features, skin-based TENGs are developed to transform physical parameters such as pressure, sliding, and other physiological variables into electronic signals, which exploit potential application. For realizing visual-image recognition, a self-powered brain-linked vision electronic-skin (e-skin) for mimicking retina is achieved from polypyrrole/

polydimethysiloxane (ppy/PDMS) triboelectric-photodetecting pixel-addressable matrix [48]. The e-skin can directly transmit photodetecting signal into brain for participating in the vision perception and behavioral intervention. Besides visual-image recognitio, more functional sensors including sliding sensor [49], touch screen [50], pressing sensor [51], and motion sensors [52] are also deeply explored.

In order to satisfy the requirement of self-powered, highly stretchability, and transparency of triboelectric skins, different materials including silicone rubber [53], metal nanowire [54, 55], and conductive polymer [56] are widely studied. To introduce the characteristic of instilling self-healing and further enhance the performance of energy generation, stretch ability, transparency, and slime-based ionic conductors were first used as transparent current-collecting layers of TENG for harvesting mechanical [57]. The ionic-skin TENG consists of a silicone rubber layer with a thickness of 100 ± 10 μm, utilized as the triboelectrically negative material, a slime layer (a crosslinked poly(vinyl alcohol) gel) with a thickness of 1 mm that works as the ionic current collector, and a VHB tape with a thickness of 1 mm as the substrate (**Figure 3a**). **Figure 3b** shows the photograph of the real highly transparent ionic-skin TENG. As depicted in **Figure 3c**, the resulting ionic-skin TENG displays a transparency of 92% transmittance for visible light. The mechanism of the ionic-skin TENG is based on the single-electrode mode, wherein human skin and silicone rubber serve as frictional layer, respectively (**Figure 3d**). **Figure 3e** shows the digital photographs of the fabricated ionic-skin TENG suffering various mechanical deformations including uniaxial stretching up to 700% strain as well as folding and rolling. The produced slime exhibits high ionic conductivity due to the presence of positive (Na^+) and negative ions ($B(OH)^{4-}$), which is measured using electrochemical impedance spectroscopy (**Figure 3f**). Thanks for the series of design, the energy-harvesting performance of ionic-skin TENG is 12-fold higher than that of the silver-based electronic current collectors. Besides, fabricated ionic-skin TENG can recover its property even suffering 300 times of complete bifurcation, exhibiting an autonomously self-healing capacity.

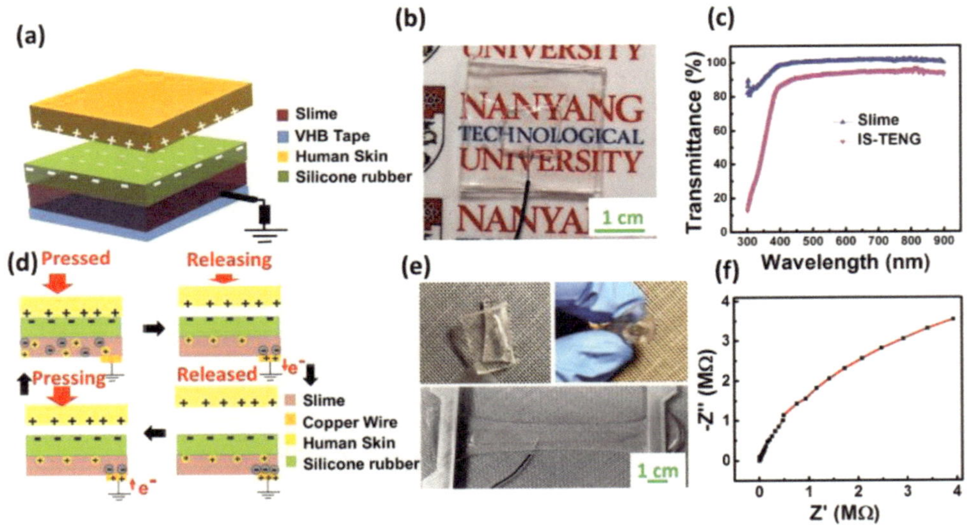

Figure 3.
(a) Schematic diagram of the IS-TENG. (b) Digital photo of the highly transparent IS-TENG. (c) Transmittance spectra of the slime (ionic conductor) and the IS-TENG with respect to a glass slide. (d) Schematic illustration of the working mechanism of the IS-TENG. (e) Digital photos of the IS-TENG under various mechanically deformed states such as axial strain up to 700%, rolled, and folded. (f) EIS measurement of the slime (ionic conductor) [57].

For versatile scavenging mechanical energy induced from arbitrary mechanical moving objects such as humans, a new mode of triboelectric nanogenerator is first demonstrated based on the sliding of a freestanding triboelectric-layer between two stationary electrodes on the same plane [58]. With two electrodes alternatively approached by the tribo-charges on the sliding layer, electricity is effectively generated due to electrostatic induction. To reduce the direct friction between triboelectric layers for energy loss, a linear grating-structured freestanding triboelectric-layer nanogenerator (GF-TENG), consisting of a freestanding triboelectric layer with grating segments and two interdigitated metal electrodes, was developed for high-efficiency harvesting vibration energy from human walking [59]. As shown in **Figure 4a**, 60 commercial LEDs (Nichia NSPG500DS) can be lighted up instantaneously with the motion of hand sliding under a slow speed and a small displacement. The GF-TENG can also havest energy from the monement of car for powering electronic components on the vehicle (**Figure 4b**). Four identical extension springs are used to suspend and anchore the triboelectric layer, as displayed in **Figure 4c**. Owing to the structure, the obtained GF-TENG can scavenge

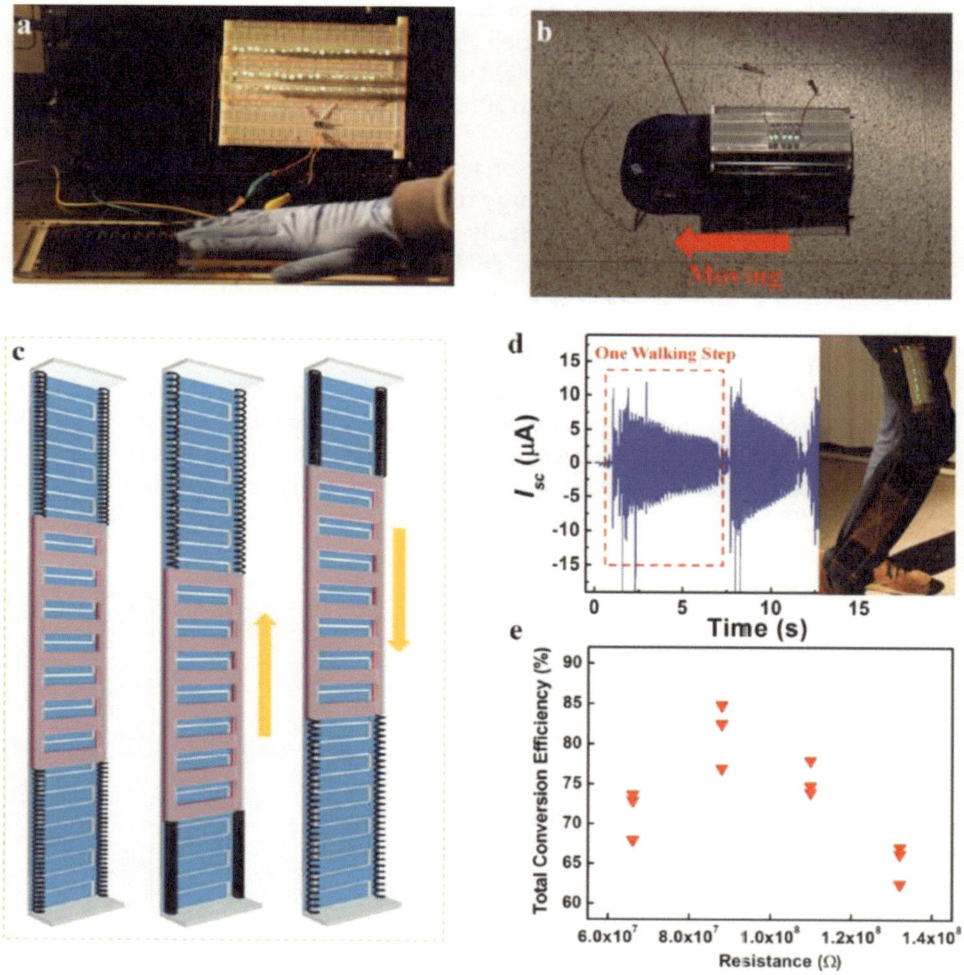

Figure 4.
Applications of GF-TENG for harvesting a wide range of mechanical energy. (a) Harvesting energy from sliding of a human hand. (b) Harvesting energy from acceleration or deceleration of a remote control car. (c) Device structure for noncontact GF-TENG. (d) Harvesting energy from people walking by noncontact GF-TENG and the real-time measurement of Isc. (e) Total conversion efficiency of noncontact GF-TENG for harvesting slight vibration under different load resistances [59].

the mechanical energy from people's walking motion when it is bonded to human legs (**Figure 4d**). An excellent stability and maxmiun energy conversion efficiency of 85% are realized at a matched load resistance of 88 MU under the noncontact mode (**Figure 4e**).

3.2 TENGs harvesting energy from vibration

Vibration, as a type of common mechanical phenomena, ubiquitously exists in ambient environment in a variety of forms and wide range of scales. Therefore, vibration can be regarded as a sustainable source of power for driving small electronics if it can be effectively collected. Contributing to the distinctive working mechanism, TENG has been proposed recently and proved a promising approach for scavenging mechanical energy from vibration, especially in the low-frequency range. To date, a variety of device and machine-based TENGs have been applied to convert mechanical energy induced from vibration into electric energy.

Chen et al. presented a harmonic-resonator-based TENG as a sustainable power source and an active vibration sensor [60]. The harmonic-resonator-based TENG, held a multilayer structure consisting of aluminum with nanoporous surface as contact electrode and nanowire-modified PTFE as frictional layer, is the first TENG that can harness random and tiny ambient vibration. It can effectively respond to vibration frequencies ranging from 2 to 200 Hz with a considerably wide working bandwidth of 13.4 Hz.

The above-mentioned harmonic resonator-based TENG with a simple structure design can only scavenge vibration energy from a single direction. In practice, vibrations in living environments generally display multiple motion directions. With this in mind, a three-dimensional TENG (3D-TENG) was designed for harvesting random vibration energy from multiple directions [61]. The 3D-TENG has a multilayer structure with circular acrylic as supporting substrates, as shown schematically in **Figure 5a**. The cylindroid core of the 3D-TENG lies at the center of the acrylic substrate with a bottom diameter of 3 cm. On the top of the core, an iron mass is mobile and suspended by three identical springs with an included angle of 120° between each other. The designed structural symmetry ensures that the whole system has a constant resonant frequency in arbitrary in-plane directions. A layer of PTFE film as one contact surface is adhered onto the bottom side of

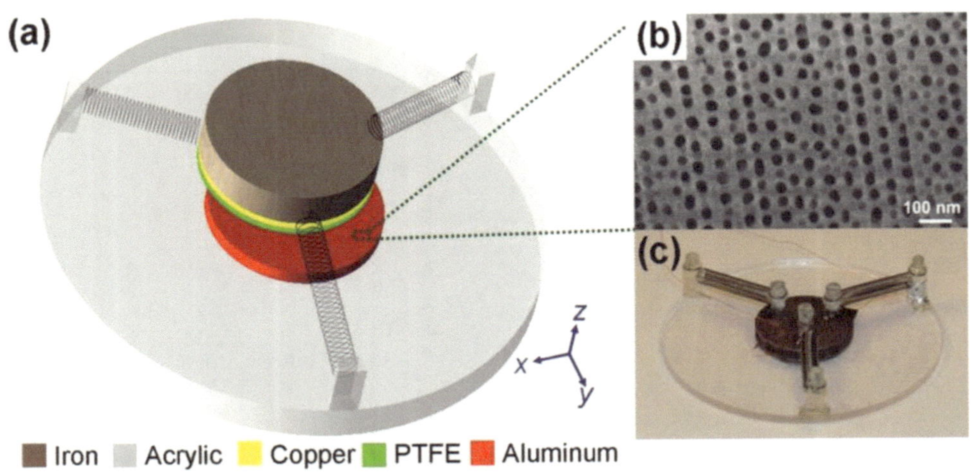

Figure 5.
3D triboelectric nanogenerator: (a) schematic of a 3D-TENG, (b) SEM image of nanopores on an aluminum electrode, and (c) a photograph of the fabricated 3D-TENG [61].

circular iron mass with a deposited copper thin film as the back electrode. Attached to the bottom acrylic substrate, an aluminum thin film with nanopore modification plays dual roles as a contact electrode and the other contact surface. The scanning electron microscopy (SEM) images of aluminum nanopores can be observed in **Figure 5b**. A photograph of the real 3D-TENG device is shown in **Figure 5c**. Owing to the conical-shaped spring structure, the 3D-TENG can operate in a hybridization mode combining with the vertical contact-separation mode and the in-plane sliding mode, which is beneficial to harvest random vibrational energy in multiple directions over a wide bandwidth.

For better sensitivity response to external disturbance, a suspended 3D spiral structure was integrated with a TENG for energy harvesting and sensor applications [62]. Operating in the vertical contact-separation mode, the desired TENG with unstable mechanical structure can balance itself when be oscillated, which makes it a superior choice for vibration energy harvesting and vibration detection. The newly designed TENG has a wide working bandwidth of 30 Hz in low-frequency range with a maximum output power density of 2.76 W/m^2 on a load of 6 MΩ.

Beyond that, a spherical three-dimensional TENG (3D-TENG) with a single electrode, consisting of an outer transparent shell and an inner polyfluoroalkoxy (PFA) ball, was designed for scavenging ambient vibration energy in full space [63]. By working at a hybridization of both the contact-separation mode and the sliding mode, the 3D-TENG can deliver a maximal output voltage of 57 V, a maximal output current of 2.3 μA, and a corresponding output power of 128 μW on a load of 100 MΩ, which can be used to directly drive tens of green light-emitting diodes. Moreover, the TENG is utilized to design the self-powered acceleration sensor with a detection sensitivity of 15.56 V/g.

Besides multiple motion directions, ambient vibrations generally exhibit a wide spectrum of frequency distribution. To solve this problem, a TENG with a wavy-structured Cu-Kapton-Cu sandwiched between two flat nanostructured PTFE films was designed for broadband vibration energy harvesting [64]. The core of the wavy structure is composed of a set of metal rods (with a diameter of 1/4 in.), as shown in **Figure 6a**. PTFE films are processed with inductively coupled plasma (ICP) etching to produce the nanostructures shown in **Figure 6b**, which would largely enhance contact electrification. The device structure is schematically shown in **Figure 6c**, accompanied by a magnified schematic in **Figure 6d** and a picture of a real device in **Figure 6e**. This structure design allows the TENG to be self-restorable after impact without the use of extra springs and converts direct impact into lateral sliding. Based on the wavy structure, the TENG can harvest vibrational energy from 5 to 500 Hz, and the generator's resonance frequency was determined to be ~100 Hz at a broad full width at half-maximum of over 100 Hz, producing a Voc of up to 72 V, an Isc of up to 32 μA, and a peak power density of 0.4 W/m^2.

After that, an elastic multiunit TENG was also realized to efficiently harvest low-frequency vibration energy over a wide frequency range [65]. The obtained TENG can provide a maximum instantaneous output power density of 102 W/m^3 at as low as 7 Hz and maintain its stable current outputs over a wide frequency range (from 5 to 25 Hz). Besides, it can act as an active vibration sensor to monitor the running status of equipment. Moreover, by combining the TENG with a power management unit to form a self-charging power unit, the vibration energy harvesting from ambient environment, such as an operating machine and running bicycle, can sustain power electronics such as thermometers, humidity sensors, speedometers, and a micro-meteorological instrument.

For improving the lower output current, a multi-layered stacked TENG was reported as a cost-effective, simple, and robust approach for harvesting ambient vibration energy [66]. The 3D-TENG has a multilayered structure with acrylic as

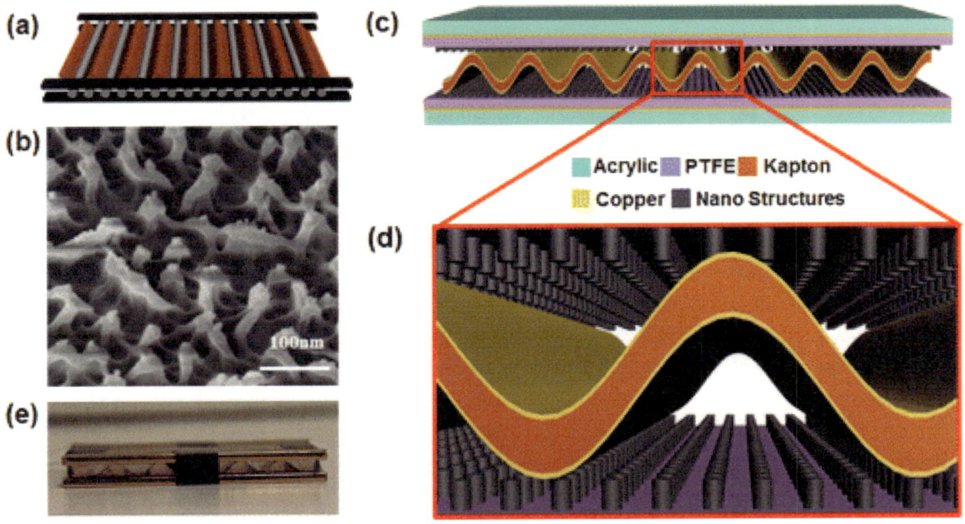

Figure 6.
(a) Schematic of the method to fabricate wavy Kapton films. (b) SEM image of the ICP-processed PTFE film surface. (c) Schematic of the device structure. (d) Magnified schematic of the device, showing that the wavy core is in periodical contact with the nanostructures on the PTFE films. (e) Photograph of an as-fabricated TENG device before packaging [64].

supporting substrates, as schematically shown in **Figure 7a**. A photograph of an as-fabricated TENG and SEM image of the PTFE nanowires is shown in **Figure 7b-c**. With superior synchronization, the 3D-TENG produces a short-circuit current as high as 1.14 mA and an Voc up to 303 V with a remarkable peak power density of 104.6 W/m^2. As a direct power source, it is capable of simultaneously lighting up 20 spot lights as well as a white G16 globe light.

To reduce the direct friction between triboelectric layers, a liquid-metal-based TENG (LM-TENG) was developed for high-efficiency vibration energy harvesting [67]. Owing to an intimate contact between the liquid metal and the polymer dielectric layer, the direct friction between triboelectric layers for energy loss is effectively reduced, resulting in high effective contact, shape adaptability, and low friction coefficient with solid. Therefore, the LM-TENG exhibits an output charge density of 430 μC/m^2, which is four to five times higher than that in the case if the electrode is solid film.

On the other hand, soft electrode can effectively increase the contact intimacy between the triboelectric layers [68]. Xu et al. reported a novel soft and robust TENG made of a silicone rubber-spring helical structure with nanocomposite-based elastomeric electrodes for harvesting arbitrary directional vibration energy and self-powered vibration sensing [69]. The schematic diagram and a photo of the S-TENG are shown in **Figure 8a,c**, respectively. As displayed, the TENG exhibits a helical structure based on the integration of elastomer and spring. A mixing well silicone rubber and carbon nanofiber, which can be stretched up to the strain of 133%, serves as the elastomeric electrode (**Figure 8b**). The working mechanisms of the S-TENG under vertical and horizontal vibration are shown in **Figure 8d,e**, respectively. Under external vertical vibration excitation, the distance between a helical structure's adjacent surfaces changes, forming a contact-separation mode TENG. Under horizontal vibration excitation, the S-TENG's helical structure's adjacent surfaces can contact on one side and separate on the other side, also forming a contact-separation mode TENG. Under the resonant states of the S-TENG, its peak power density is found to be 240 and 45 mW/m^2 with an external load of

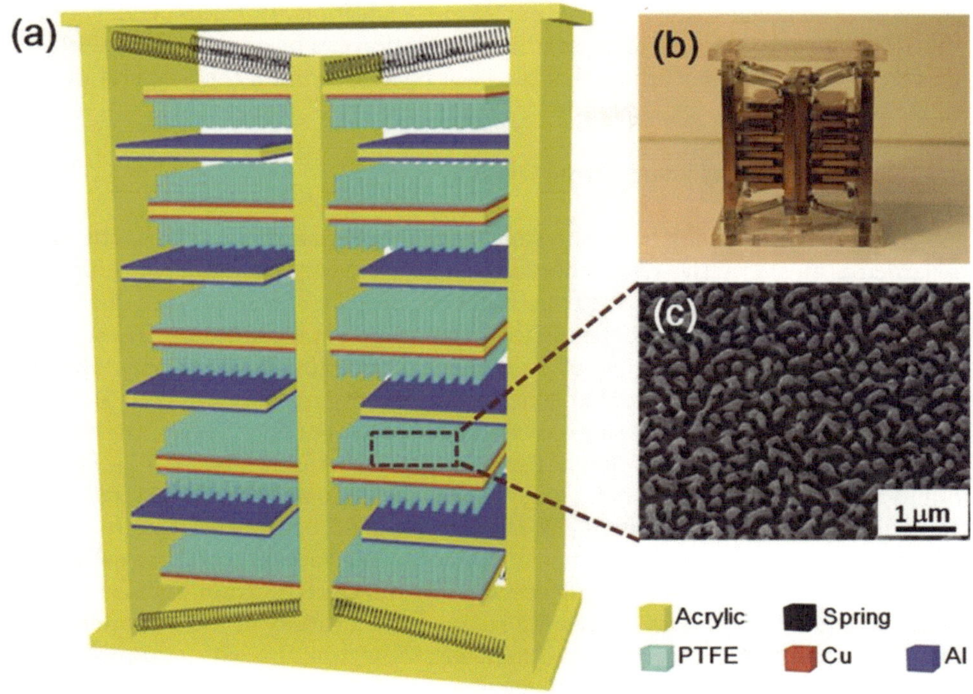

Figure 7.
Three-dimensional triboelectric nanogenerator. (a) Schematic of a 3D-TENG. (b) SEM image of nanopores on aluminum electrode. (c) A photograph of the as-fabricated 3D-TENG [66].

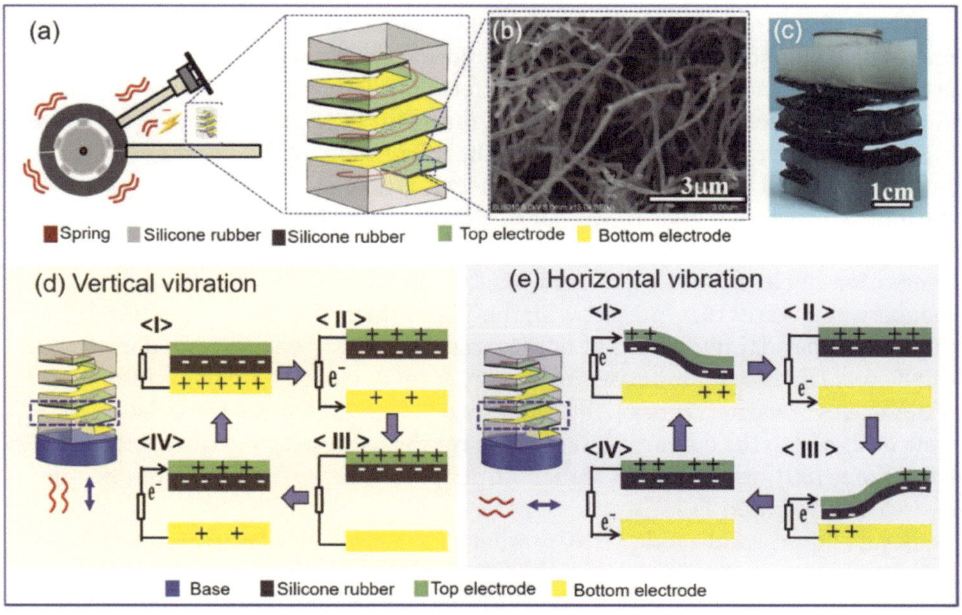

Figure 8.
(a) The device schematic of the S-TENG. Note that the gray silicone rubber layer containing a spring forms a base on which other layers can be built, and the black silicone rubber layer along with the electrode layer forms a contact-separation pair. Both top and bottom electrodes are made of carbon nanofiber-mixed silicone rubber. (b) SEM image of the carbon nanofiber for preparing the elastomeric electrode. (c) Photo of the as-prepared S-TENG. Working mechanisms of the S-TENG under (d) vertical vibration excitation and (e) horizontal vibration excitation [69].

10 MΩ and an acceleration amplitude of 23 m/s^2. Additionally, the dependence of the S-TENG's output signal on the ambient excitation can be used as a prime self-powered active vibration sensor that can be applied to monitor the acceleration and frequency of the ambient excitation.

3.3 TENGs harvesting energy from water

Water energy deriving from rainwater, ocean waves, and waterfalls has been regarded as an alternative renewable energy resource source without polluting the environment. Energy harvesting from water has been further reinforced due to the abundant reserves and little dependence on environmental conditions. Through decades of exploration, a variety of wave energy converting devices and machines based on TENG has been invented to harvesting energy from water.

Liquid-solid-mode TENGs for harvesting liquid-wave energy have drawn much attention for the features of relatively stable output and durability [70–72]. For the liquid–solid-mode TENG, contact separation is the main representative strategies applied to scavenge water energy [73, 74]. A hydrophobic surface on water-solid TENGs is beneficial for inducing separation at the interface of liquid and solid [75]. Based on this, Zhu et al. reported a liquid-solid electrification-enabled TENG based on a FEP thin film for harvesting energy from a variety of water motions [76]. Owing to the modification of aligned nanowires, the thin film with a property of hydrophobicity can increase the contact area at the liquid-solid interface, leading to enhanced surface charging density and thus electric output at an efficiency of 7.7%. Due to the creation of continuous contact separation between water and the solid surface, a cylindrical water TENG was designed by using a hydrophilic surface along with the hydrophobic surface to control the water flow inside a packaged system for enhanced electrostatic induction [77].

Generally, an effective way of integrating a number of electrodes together to make them area scalable is helpful for promoting output power density. On the other hand, the electric power is highly affected by nanostructures at the solid/liquid interface. According to this, a flexible thin-film TENG was reported for harvesting kinetic wave energy [78]. Because of the integration method that use an array of surface-mounted bridge rectifiers to connect multiple parallel electrode together, the induced current between any pair of electrodes can be constructively added up, leading to a significant enhancement in output power and realizing area-scalable integration of electrode arrays. However, the thin-film TENG is only applicable to regular water waves that interact with the TENG through a linear water level. For improving the adaptive means of harvesting water energy, a networked integrated TENG was fabricated for harvesting energy from interfacing interactions with water waves of various types [79]. Additionally, interdigital electrode-based TENGs were designed in the contact-sliding mode for the harvesting of triboelectric energy from water [80], resulting in a higher output performance than those of one- and two-electrode-based TENGs.

Beside liquid-solid-mode TENGs, other structure TENGs were designed for harvesting water energy generating by flowing water, such as multi-layered disk structure [81], floating buoy structure [82], radial-arrayed rotary structure [10], and so on. Although many water-based TENGs have been fabricated, there is a lack of effort in realizing TENG harvesting water energy directly on the fabric/textile, due to the poor water resistance of the fabrics related to their intrinsic hydrophilicity that can be ascribed to their abundant hydrophilic groups, and the strong adsorption capacity because of their large specific surface area [83]. For realizing the practical wearable device harvesting energy from water flow, Xiong et al. reported a wearable fabric-based WTEG with additional self-cleaning and antifouling performance for the first time [83]. This is realized with the preparation of hydrophobic cellulose oleoyl ester nanoparticles by a nontoxic esterification method and nanoprecipitation technology based on the microcrystalline cellulose. In this study, PET fabric-based WTEG can generate the output power density of 0.14 W/m^2 at a load resistance of 100 MΩ.

There are two parts to water wave energy including the electrostatic energy from the contact electrification between water and surrounding media and the mechanical impact energy. For simultaneously scavenging both the energy from water, some works have been well done. For example, Su et al. presented an all-in-one hybridized TENG based on the conjunction of liquid-solid interfacial electrification enabled TENG and impact-TENG for harvesting water wave energy and as a self-powered distress signal emitter [84]; Lin et al. designed a fully integrated TENG for harvesting water energy and as a self-powered ethanol nanosensor, which contained a water-TENG unit to collect the electrostatic energy of water and a contact-TENG unit to collect the mechanical/kinetic energy of water [85]; Cheng et al. developed a water wheel hybridized TENG, composed of a water-TENG part and a disk-TENG part, for simultaneously harvesting the two types of energies from the tap water flowing from a household faucet [86]. Based on a unique structure design, the hybridized TENGs are shown to be suitable for harvesting multiple types of energies from water.

During a working process, the acting surfaces of the above mentioned TENGs will be exposed to ambient atmosphere, which will limit their applications in some cases. The interface electrification was seriously affected by humidity, causing a quick decline of the surface charge density [87]. In order to improve the performance of TENGs under harsh conditions with the presence of water, fully enclosed or packaged TENGs should be developed for tolerating the environment. So far, different designs were developed based on packaged TENG such as wavy-shaped models [88], fully packaged contact-separation configurations [89–91], and rolling spherical structure [92]. Wang et al. designed a freestanding, fully enclosed TENG that encloses a rolling ball inside a rocking spherical shell for harvesting low-frequency water wave energy [93]. An image of the fabricated TENG floating on water is shown in **Figure 9a**. **Figure 9b** shows the schematic diagram of the freestanding structured design that consists of one rolling ball and two stationary electrodes. To enhance the electric output of the TENG, nanowire arrays are fabricated on the surface of the Kapton film (**Figure 9c**) that provides a large contact area to generate more triboelectric charges on the surface. Through the optimization of materials and structural parameters, a spherical TENG of 6 cm in diameter actuated by water waves can provide a peak current of 1 μA over a wide load range from a short-circuit condition to 10 GΩ, with an instantaneous output power of up to 10 mW. This rolling-structured TENG is extremely lightweight, has a simple structure, and is capable of rocking

Figure 9.
Device structure, basic operations of the freestanding-triboelectric-layer-based nanogenerator (RF-TENG) with a rolling Nylon ball enclosed. (a) Photograph of a rocking nanogenerator floating on water. (b) Schematic diagrams of freestanding-structured design. (c) SEM image of nanorod structure on the Kapton surface [93].

on or in water to harvest wave energy. Additionally, rolling spherical TENGs and coupled TENG networks have been demonstrated to harness the water wave energy because of the advantages of light-weight, small-resistance under the water wave motions, and easy to be integrated [94, 95].

For enhancing the output current and enlarging the practical applications of packaged TENG, introducing a spring structure into the TENG can store the kinetic energy from water impact and later convert into electric power via residual vibrations [96]. Combining the advantages of spring structure and integrated multilayered structure, Xiao et al. demonstrated a kind of spherical TENG with spring-assisted multilayered structure for harvesting water wave energy [97]. The introduction of spring structure enhances the output performance of the spherical TENG by transforming low-frequency water wave motions into high-frequency vibrations, while the multilayered structure increases the space utilization, leading to a higher output of a spherical unit. The structure of spherical TENG designed with spring-assisted multilayered structure floating on water surface is schematically shown in **Figure 10a**. **Figure 10b** displays a photograph of as-fabricated spherical TENG device, and the inset shows the photograph of the device in the water waves. The working principle of each TENG unit is demonstrated in **Figure 10c**. The periodic movement of the mass block under the triggering of water waves, which leads to the contact and separation between two surfaces of the top aluminum foil and FEP film, produces periodic electric output signals. Owing to its unique structure, the output current of one spherical TENG unit can reach 120 μA, which is two orders of magnitude larger than that of previous rolling spherical TENG, and a maximum output power up to 7.96 mW is realized as triggered by the water waves.

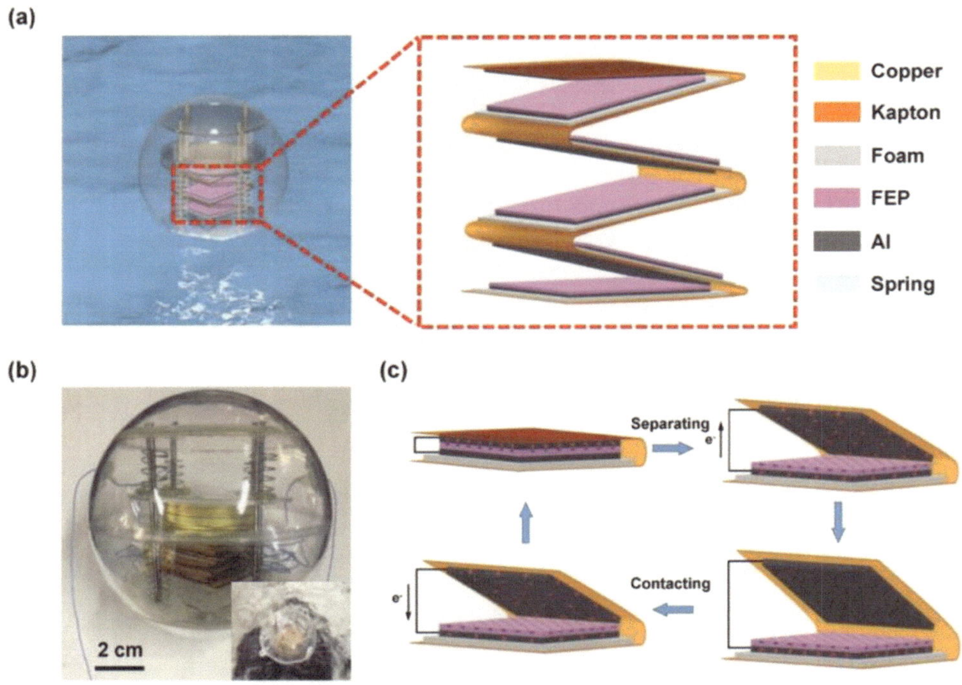

Figure 10.
(a) Schematic diagram of the spherical TENG with spring-assisted multilayered structure floating on water, and schematic representation enlarged structure for the zigzag multilayered TENG with five basic units. (b) Photographs of the as-fabricated TENG device. (c) Working principle of each TENG unit of the spherical TENG [97].

3.4 TENGs harvesting energy from wind

Wind energy can be a renewable energy sources for energy harvesting on account of widespread and absolute abundance. The practical application of traditional wind power in our daily life is largely limited by the extra-large volume, high cost of installation, noise and geographical environment. In this regard, TENG is one of the most alternative wind energy conversion strategies on accord of its small scale, low cost, simple fabrication routes, and portability [98]. In order to harvest wind energy, flutter-driven structure [99, 100] and rotational structure [101, 102] are the two main methods for preparing wind-driven TENG.

Flutter-driven structure TENG for harvesting wind energy was realized by Yang et al. for the first time [103]. As displayed in **Figure 11**, the TNEG is composed of two layers of Al foils and a FEP film laying in midair of a cuboid acrylic tube. The Al foils act as both triboelectric surfaces and electrodes, respectively. The FEP film is fixed one side, leaving the other side freestanding. The FEP film will vibrate periodically to contact the two Al foils inducing from wind, resulting in an output signal in an external circuit. Output voltage and current about 100 V and 1.6 μA are achieved, and a corresponding output power of 0.16 mW is realized under a loading resistance of 100 MΩ.

Although single-side fixed-based TENG exhibits good performance for scavenging wind energy, the stability of output performance is a challenge because of the arbitrary fluttering of the FEP film. For solving the problem, an elasto-aerodynamics-driven TENG, consisting of a Kapton film with two Cu electrodes fixed on two ends in an acrylic fluid channel, was reported for scavenging air-flow energy [104], where the flutter effect of Cu electrodes was induced to contact two triboelectric materials of the PTFE films and the Kapton film to realize the output performance of the device.

Based on flutter-driven structure, many other efforts have been made to enhance the performance of TENG through optimizing the structure or the morphologies of material surface design. A lightweight and freestanding flag-type woven TENG, consisting of conductive belts of Ni-coated polyester textiles and Kapton film-sandwiched Cu belts, was designed for scavenging high-altitude wind energy from arbitrary directions [105]. When wind fluttering is applied in each woven unit, wind energy converts into electrical energy induced by the interlaced interactions between the Kapton film and a conductive cloth under wind-introduced fluttering of the flag. Besides, a flutter-driven TENG, consisting of a flag and a counter plate arranged in parallel with interwoven microstructure, was fabricated for harvesting wind energy based on contact electrification caused by the self-sustained oscillation of flags [106]. As shown in **Figure 12**, a flexible flag and a rigid plate are arranged in face to face in order to prepare a wind-driven energy-harvesting system using fluttering behavior. Owing to the design, interaction between them can lead to a rapid

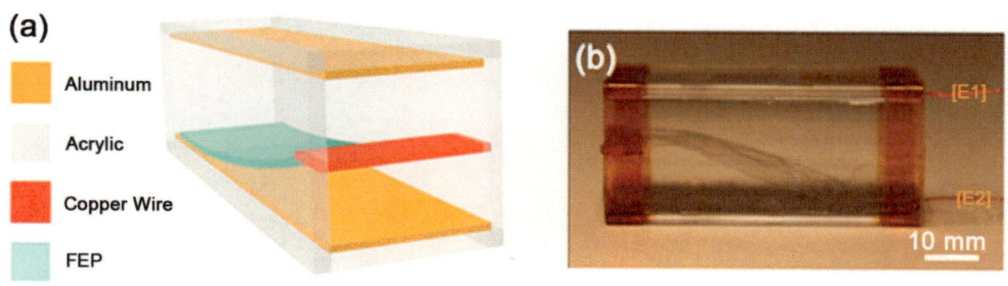

Figure 11.
(a and b) The structure and photograph of the first reported flutter-driven mode WD-TENG [103].

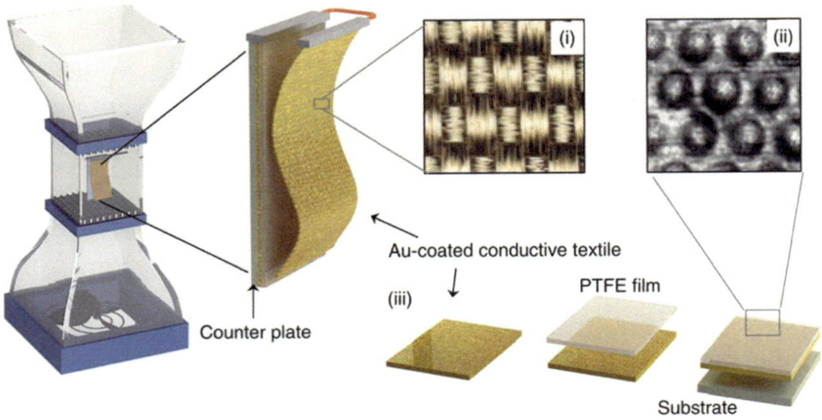

Figure 12.
Schematic diagrams of a wind tunnel and the structural design of a flutter-driven triboelectric generator including surface characteristics of (i) a highly flexible flag, (ii) a counter plate, and (iii) the fabrication of the counter plate [106].

periodic contact and separation, and that movement can be successfully employed for converting the kinetic energy of the wind into electrical energy.

For rotational structure, wind cup is a main method for scavenging wind energy. Deriving from the conventional wind cup structure, a rotary structured TENG was presented for scavenging weak wind energy in our environment [101]. As illustrated in **Figure 13**, the rotary structured TENG is composed of a framework, a shaft, a flexible rotor blade, and two stators. When wind flowing is utilized in the rotation of the shaft and the flexible rotor, a flexible and soft polyester (PET) rotor blade with a PTFE film adhered at the end will periodically sweep across the Al electrodes. In this process, a consecutive face-to-face contact and separation between PTFE film and Al electrodes are produced, regarding as the basic process for generating electricity.

Aiming to improve the robustness and lifetime of wind-driven TENG, a freestanding disk-based TENG was fabricated to harvest wind energy through automatic transition between contact and noncontact working states [102]. The major structure of the disk-based TENG includes two parts: the rotational inner acrylic barrel that connects with the freestanding rotor of the disk TENG and the stationary outer barrel that connects with the stator of the TENG. Two bearings

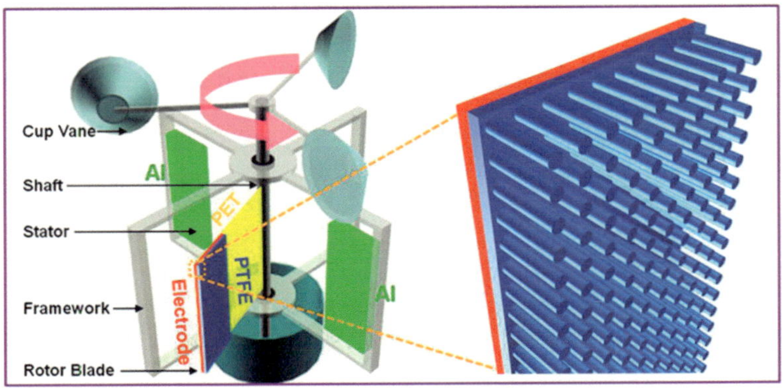

Figure 13.
The schematic diagram showing the structural design of the R-TENG, with the enlarged picture showing the nanowire-like structures on the surface of PTFE [101].

are used to link the two parts and enable the relative rotation. Benefiting from the unique structural design, the TENG can work in the noncontact state with minimum surface wear and also transit into contact state intermittently to maintain high triboelectric charge density.

Besides serving as a power source for running some electric devices, wind-driven TENG is also expected to be utilized as various self-power systems by integrating with other electric devices. Chen et al. introduced the first self-powered air cleaning system focusing on sulfur dioxide (SO_2) and dust removal as driven by the electricity generated by natural wind, with the use of rotating TENG [107]. Another common wind-driven TENG-based self-power system is the wind speed sensor. Kim et al. prepared wind-driven TENG based on rolling motion of beads for harvesting wind energy as a self-power wind speed sensor [108]. Wen et al. fabricated a blow-driven TENG, acting as an active alcohol breath analyzer, which is featured as high detection gas response of ~34 under an optimized sensor working temperature, fast response time of 11 s as well as a fast recovery of 20 s [109].

3.5 Hybrid nanogenerator

Aiming to simultaneously harvesting multitypes of energies from various sources, TENG has been hybridized with various other energy harvester strategies from the environment. It is well known that solar irradiance is another clean and renewable energy sources. To develop a practical method to simultaneously scavenge solar and mechanical energies, the concept of a hybridized energy harvester integrating TENG and solar cell was presented [110, 111]. Based on lightweight and low cost, fabric-based material is served as the ideal strategy utilized to fabricate these kinds of hybrid generator [112]. Chen et al. presented a foldable and sustainable power source by fabricating an all-solid hybrid power textile with economically viable materials and scalable fabrication technologies [34]. The wearable all-solid hybrid power textile has a single-layer interlaced structure, which is a mixture of two polymer-wire-based energy harvesters, including both a fabric TENG to convert mechanical movement into electricity and a photovoltaic textile to gather power from ambient sunlight, as schematically illustrated in **Figure 14a,b**, respectively. An enlarged view of the interlaced structure is presented for both the fabric TENG (**Figure 14c**) and photovoltaic textile (**Figure 14d**). Under ambient sunlight with mechanical excitation, like human motion, car movement, and wind blowing, the as-woven textile was capable of generating sufficient power for various practical applications, including charging a 2 mF commercial capacitor up to 2 V in 1 min, continuously driving an electronic watch, directly charging a cell phone, and driving the water splitting reactions.

Aiming to largely collect the energy from mechanical motions, an integrated TENG and an electromagnetic generator (EMG) for concurrently harvesting mechanical energy are a promising way. By integrating two kinds of mechanical energy harvesting units, the weight of the EMG can be reduced and the total output power can be increased to expand the potential applications [113–117]. In them, rotational structure is the typical strategy utilized to simultaneously convert mechanical energy into electrical energy from one rotating motion. By integrating an EMG and a TENG, a rotation-based hybrid generator is first fabricated to generate a high output that can sustainably drive a commercial globe light with an intensity of illumination up to 1700 lx [118]. As illustrated in **Figure 15a**, the main structure of the hybrid generator consists of an EMG including the top and bottom layers (1 and 5) and a TENG including the middle layers (2, 3, and 4) with the planar structures, where the rotator and the stator are composed of layers 1 and2 and layers 3–5, respectively. The corresponding photographs of each layer are displayed

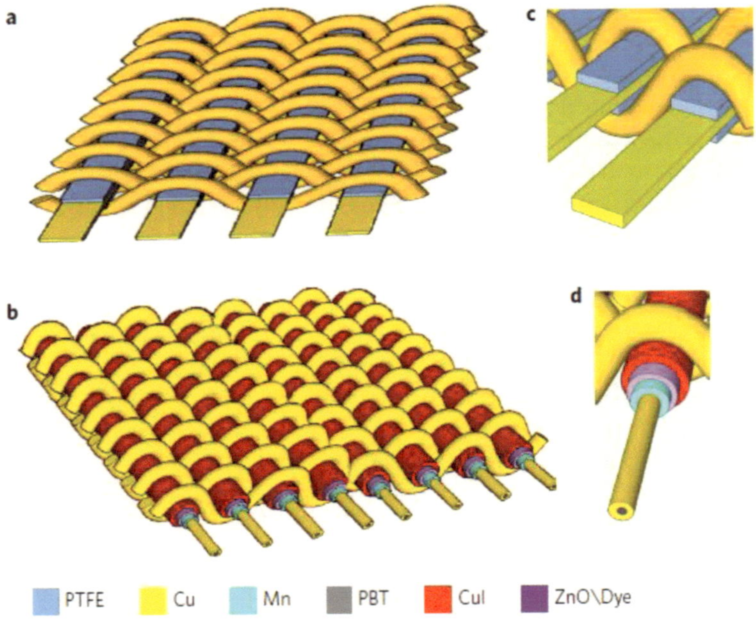

Figure 14.
Structural design of the hybrid power textile. (a and b) Schematic illustration of the hybrid power textile, which is a mixture of two textile-based all-solid energy harvesters: fabric TENG (a) and photovoltaic textile(b). Enlarged view of the interlaced structure of both the fabric TENG (c) and the photovoltaic textile (d) [112].

in **Figure 15b**. Based on the relative rotation between the rotator and the stator, the hybrid generator simultaneously collects biomechanical energy from human hand-induced rotating motions. In order to compare the two generators with each other systematically, Guo et al. fabricated a water-proof triboelectric-electromagnetic hybrid generator, including a fully enclosed packaging of TENG achieved by the interactions between pairs of magnets as the noncontact mechanical transmission forces [119]. Systematic study of the influences of the designed parameters, including the segment's number of the TENG, the rotation speed, and the arrangement of the coils, on the electrical outputs of the WPHG were performed experimentally. The result demonstrated that TENG can produce a stable voltage to power commercial electronic device even under a low rotation speed compared with EMG.

Besides the above mentioned, other strategies have been applied to intergrate with TENG for collecting other types of energies. Lee et al. presented a flexible hybrid cell to simultaneously harvest thermal and mechanical energies from skin temperature and body motion [120]. For fabricating the hybrid cell, ZnO nanowires are grown on the sputtered-coated seed layer surface of a thin Al substrate. And then, a 2-μm thick poly(methyl methacrylate) (PMMA) layer is coated on the surface of the as-grown ZnO nanowires, and a thin Al substrate is stacked on the PMMA-coated layer to be used as the top electrode. Owing to the structure design, the hybrid cell can simultaneously harvest thermal and mechanical energies so that the energy resources can be effectively and complementarily utilized for power sensor network and micro/nanosystems. Addtionally, combining the TENG with piezoelectric nanogenerator (PENG) is a alternative manner for concurrently collecting mechnical energy. Guo et al. developed an all-fiber hybrid piezoelectric-enhanced TENG that fabricated by electrospinning silk fibroin and poly(vinylidene fluoride) (PVDF) nanofibers on conductive fabrics [121]. Contributing to the large specific surface area of nanofibers and the extraordinary ability of silk fibroin to donate electrons in triboelectrification, the hybrid nanogenerator exhibited an

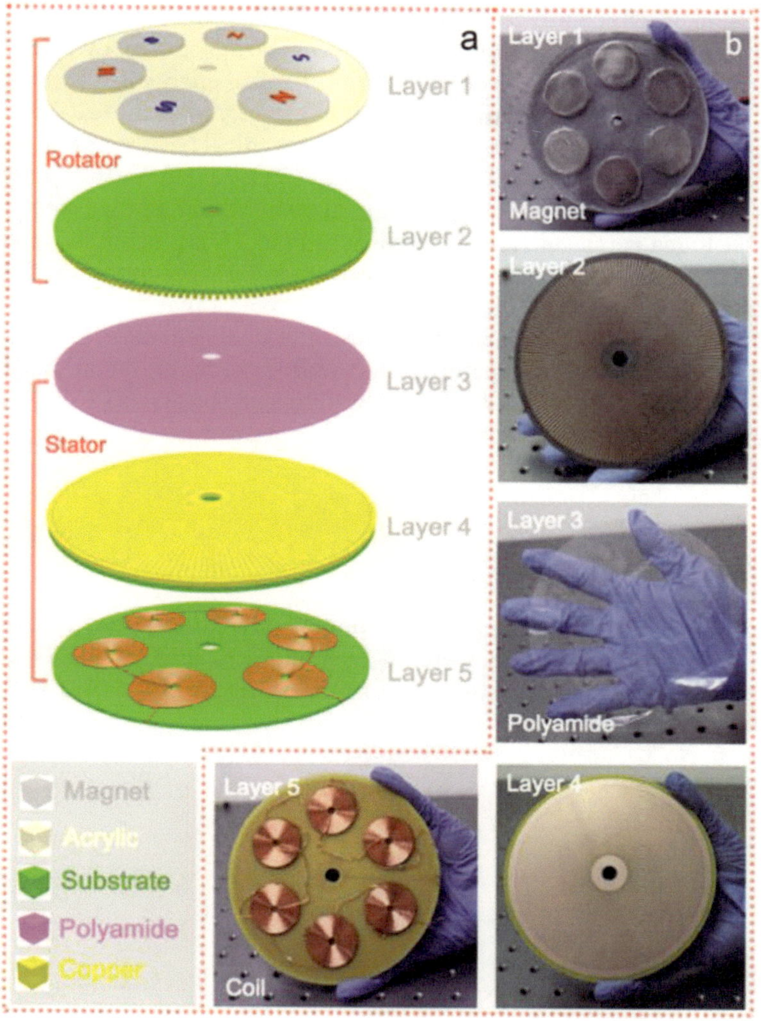

Figure 15.
(a) Schematic diagram of the designed hybridized nanogenerator. (b) Photographs of the hybridized nanogenerator [118].

outstanding electrical performance, with a power density of 310 μW/cm^2, so that it can be regarded as a self-powered wearable microsystem for falling-down detection and timely remote alarm.

4. Conclusions

In order to seek an intelligent life, trillions of electronic device for the Internet of Things are requisite with higher personal, portable, complex, multifunctional, and smart. Aiming to maintain the normal working status of these small electronic devices sustainably, an effective technology to harvest small-scale energy from renewable natural resources is highly desirable. Given the collection characteristics of simple structure, flexibility, low cost, light weight, high efficiency, high power density, and environmental friendly, the invention of TENG is served as an promising small-scale energy harvester who can convert mechanical motions into electricity, even at low frequency. Futhermore, TENGs can also be utilized to transform physical parameters such as pressure, sliding, and other physiological variables into electronic signals, which directly reflected the information of mechanical stimuli and environmental conditions without an external power source. By extensively investigating, TENG can effectively harvest mechanical energy in almost any form based on the four fundamental modes, and thus can regard as the self-powered sensors for a wide application under diffident mechanical triggerings. In the future, the continuous endeavors on TENGs will lagerly enhance their output performance. Based on deeply investigating the fundamental menchanism of triboelectrification, it is possible to realize the ultrahigh charge density of TENG via material modification, structure design, or condition optimization. Besides the output perfoermance, the durability and output stability is the other bottleneck that limited the application of TENG, especially comparing with the traditional generator. It might overcome through fabricating new materials or coupling modes of operations. Based on the above discussion and analysis, it can be anticipated that TENG will soon become an ideal small-scall energy haverter with broad application as self-powered sensors through the world wide efforts.

Acknowledgements

The authors like to thank the financial supports from the National Key R & D Project from Minister of Science and Technology (2016YFA0202704), Beijing Municipal Science & Technology Commission (Z171100000317001, Z171100002017017, Y3993113DF), and National Natural Science Foundation of China (Grant No. 61774016, 21773009, 51432005, 5151101243, 51561145021).

Conflict of interest

There is no conflict of interest.

Author details

Jie Wang[1,2]*, Linglin Zhou[1,2], Chunlei Zhang[1,2] and Zhong Lin Wang[1,2,3]

1 Beijing Institute of Nanoenergy and Nanosystems, Chinese Academy of Sciences, Beijing, PR China

2 College of Nanoscience and Technology, University of Chinese Academy of Sciences, Beijing, PR China

3 School of Materials Science and Engineering, Georgia Institute of Technology, Atlanta, GA, USA

*Address all correspondence to: wangjie@binn.cas.cn

References

[1] Atzori L, Iera A, Morabito G. The Internet of Things: A survey. Computer Networks. 2010;**54**:2787-2805. DOI: 10.1016/j.comnet.2010.05.010

[2] Miorandi D, Sicari S, De Pellegrini F, Chlamtac I. Internet of Things: Vision, applications and research challenges. Ad Hoc Networks. 2012;**10**:1497-1516. DOI: 10.1016/j.adhoc.2012.02.016

[3] Wu CS, Ding WB, Liu RY, Wang JY, Wang AC, Wang J, et al. Keystroke dynamics enabled authentication and identification using triboelectric nanogenerator array. Materials Today. 2018;**21**:216-222. DOI: 10.1016/j.mattod.2018.01.006

[4] Luo DY, Yang WQ, Wang ZP, Sadhanala A, Hu Q, Su R, et al. Enhanced photovoltage for inverted planar heterojunction perovskite solar cells. Science. 2018;**360**:1442-1446. DOI: 10.1126/science.aap9282

[5] Vats G, Kumar A, Ortega N, Bowen CR, Katiyar RS. Pyroelectric control of magnetization for tuning thermomagnetic energy conversion and magnetocaloric effect. Energy & Environmental Science. 2016;**9**:2383-2391. DOI: 10.1039/c6ee01013j

[6] Yang B, Lee C, Xiang WF, Xie J, He JH, Kotlanka RK, et al. Electromagnetic energy harvesting from vibrations of multiple frequencies. Journal of Micromechanics and Microengineering. 2009;**19**:8. DOI: 10.1088/0960-1317/19/3/035001

[7] Wang XD, Song JH, Liu J, Wang ZL. Direct-current nanogenerator driven by ultrasonic waves. Science. 2007;**316**:102-105. DOI: 10.1126/science.1139366

[8] Wang ZL, Song JH. Piezoelectric nanogenerators based on zinc oxide nanowire arrays. Science. 2006;**312**:242-246. DOI: 10.1126/science.1124005

[9] Wang ZL. Triboelectric nanogenerators as new energy technology and self-powered sensors—Principles, problems and perspectives. Faraday Discussions. 2014;**176**:447-458. DOI: 10.1039/c4fd00159a

[10] Zhu G, Chen J, Zhang TJ, Jing QS, Wang ZL. Radial-arrayed rotary electrification for high performance triboelectric generator. Nature Communications. 2014;**5**:9. DOI: 10.1038/ncomms4426

[11] Wang SH, Lin L, Wang ZL. Triboelectric nanogenerators as self-powered active sensors. Nano Energy. 2015;**11**:436-462. DOI: 10.1016/j.nanoen.2014.10.034

[12] Wang ZL. On Maxwell's displacement current for energy and sensors: The origin of nanogenerators. Materials Today. 2017;**20**:74-82. DOI: 10.1016/j.mattod.2016.12.001

[13] Jing QS, Xie YN, Zhu G, Han RPS, Wang ZL. Self-powered thin-film motion vector sensor. Nature Communications. 2015;**6**:8. DOI: 10.1038/ncomms9031

[14] Niu SM, Wang XF, Yi F, Zhou YS, Wang ZL. A universal self-charging system driven by random biomechanical energy for sustainable operation of mobile electronics. Nature Communications. 2015;**6**:8. DOI: 10.1038/ncomms9975

[15] Zi YL, Wang J, Wang SH, Li SM, Wen Z, Guo HY, et al. Effective energy storage from a triboelectric nanogenerator. Nature Communications. 2016;**7**:8. DOI: 10.1038/ncomms10987

[16] Wu CS, Wang X, Lin L, Guo HY, Wang ZL. Paper-based triboelectric

nanogenerators made of stretchable interlocking kirigami patterns. ACS Nano. 2016;**10**:4652-4659. DOI: 10.1021/acsnano.6b00949

[17] Lee KY, Gupta MK, Kim SW. Transparent flexible stretchable piezoelectric and triboelectric nanogenerators for powering portable electronics. Nano Energy. 2015;**14**: 139-160. DOI: 10.1016/j.nanoen.2014.11.009

[18] Chandrasekhar A, Alluri NR, Saravanakumar B, Selvarajan S, Kim SJ. Human interactive triboelectric nanogenerator as a self-powered smart seat. ACS Applied Materials & Interfaces. 2016;**8**:9692-9699. DOI: 10.1021/acsami.6b00548

[19] Zhang XL, Zheng YB, Wang DA, Rahman ZU, Zhou F. Liquid-solid contact triboelectrification and its use in self-powered nanosensor for detecting organics in water. Nano Energy. 2016;**30**:321-329. DOI: 10.1016/j.nanoen.2016.10.025

[20] Wang J, Wu CS, Dai YJ, Zhao ZH, Wang A, Zhang TJ, et al. Achieving ultrahigh triboelectric charge density for efficient energy harvesting. Nature Communications. 2017;**8**:8. DOI: 10.1038/s41467-017-00131-4

[21] Wang SH, Lin L, Wang ZL. Nanoscale triboelectric-effect-enabled energy conversion for sustainably powering portable electronics. Nano Letters. 2012;**12**:6339-6346. DOI: 10.1021/nl303573d

[22] Zhu G, Pan CF, Guo WX, Chen CY, Zhou YS, Yu RM, et al. Triboelectric-generator-driven pulse electrodeposition for micropatterning. Nano Letters. 2012;**12**:4960-4965. DOI: 10.1021/nl302560k

[23] Wang SH, Lin L, Xie YN, Jing QS, Niu SM, Wang ZL. Sliding-triboelectric nanogenerators based on in-plane charge-separation mechanism. Nano Letters. 2013;**13**:2226-2233. DOI: 10.1021/nl400738p

[24] Zhu G, Chen J, Liu Y, Bai P, Zhou YS, Jing QS, et al. Linear-grating triboelectric generator based on sliding electrification. Nano Letters. 2013;**13**:2282-2289. DOI: 10.1021/nl4008985

[25] Yang Y, Zhang HL, Chen J, Jing QS, Zhou YS, Wen XN, et al. Single-electrode-based sliding triboelectric nanogenerator for self-powered displacement vector sensor system. ACS Nano. 2013;**7**:7342-7351. DOI: 10.1021/nn403021m

[26] Niu SM, Liu Y, Wang SH, Lin L, Zhou YS, Hu YF, et al. Theoretical investigation and structural optimization of single-electrode triboelectric nanogenerators. Advanced Functional Materials. 2014;**24**: 3332-3340. DOI: 10.1002/adfm.201303799

[27] Zhu G, Bai P, Chen J, Wang ZL. Power-generating shoe insole based on triboelectric nanogenerators for self-powered consumer electronics. Nano Energy. 2013;**2**:688-692. DOI: 10.1016/j.nanoen.2013.08.002

[28] Bai P, Zhu G, Lin ZH, Jing QS, Chen J, Zhang G, et al. Integrated multi layered triboelectric nanogenerator for harvesting biomechanical energy from human motions. ACS Nano. 2013;**7**:3713-3719. DOI: 10.1021/nn4007708

[29] Yang WQ, Chen J, Zhu G, Yang J, Bai P, Su YJ, et al. Harvesting energy from the natural vibration of human walking. ACS Nano. 2013;**7**:11317-11324. DOI: 10.1021/nn405175z

[30] Shen JL, Li ZL, Yu JY, Ding B. Humidity-resisting triboelectric nanogenerator for high performance biomechanical energy harvesting.

Nano Energy. 2017;**40**:282-288. DOI: 10.1016/j.nanoen.2017.08.035

[31] Zeng W, Shu L, Li Q, Chen S, Wang F, Tao XM. Fiber-based wearable electronics: A review of materials, fabrication, devices, and applications. Advanced Materials. 2014;**26**: 5310-5336. DOI: 10.1002/adma.201400633

[32] Lee S, Ko W, Oh Y, Lee J, Baek G, Lee Y, et al. Triboelectric energy harvester based on wearable textile platforms employing various surface morphologies. Nano Energy. 2015;**12**:410-418. DOI: 10.1016/j.nanoen.2015.01.009

[33] Lai Y-C, Deng J, Zhang SL, Niu S, Guo H, Wang ZL. Single-thread-based wearable and highly stretchable triboelectric nanogenerators and their applications in cloth-based self-powered human-interactive and biomedical sensing. Advanced Functional Materials. 2017;**27**:1604462. DOI: 10.1002/adfm.201604462

[34] Chen J, Huang Y, Zhang N, Zou H, Liu R, Tao C, et al. Micro-cable structured textile for simultaneously harvesting solar and mechanical energy. Nature Energy. 2016;**1**:161138. DOI: 10.1038/nenergy.2016.138

[35] Wang J, Li XH, Zi YL, Wang SH, Li ZL, Zheng L, et al. A flexible fiber-based supercapacitor-triboelectric-nanogenerator power system for wearable electronics. Advanced Materials. 2015;**27**:4830-4836. DOI: 10.1002/adma.201501934

[36] Dong K, Wang YC, Deng JN, Dai YJ, Zhang SL, Zou HY, et al. A highly stretchable and washable all-yarn-based self-charging knitting power textile composed of fiber triboelectric nanogenerators and supercapacitors. ACS Nano. 2017;**11**:9490-9499. DOI: 10.1021/acsnano.7b05317

[37] Chai ZS, Zhang NN, Sun P, Huang Y, Zhao CX, Fang HJ, et al. Tailorable and wearable textile devices for solar energy harvesting and simultaneous storage. ACS Nano. 2016;**10**: 9201-9207. DOI: 10.1021/acsnano.6b05293

[38] Chen HM, Bai L, Li T, Zhao C, Zhang JS, Zhang N, et al. Wearable and robust triboelectric nanogenerator based on crumpled gold films. Nano Energy. 2018;**46**:73-80. DOI: 10.1016/j.nanoen.2018.01.032

[39] Kim KN, Chun J, Kim JW, Lee KY, Park JU, Kim SW, et al. Highly stretchable 2D fabrics for wearable triboelectric nanogenerator under harsh environments. ACS Nano. 2015;**9**: 6394-6400. DOI: 10.1021/acsnano.5b02010

[40] Seung W, Gupta MK, Lee KY, Shin KS, Lee JH, Kim TY, et al. Nanopatterned textile-based wearable triboelectric nanogenerator. ACS Nano. 2015;**9**:3501-3509. DOI: 10.1021/nn507221f

[41] Tian ZM, He J, Chen X, Zhang ZX, Wen T, Zhai C, et al. Performance-boosted triboelectric textile for harvesting human motion energy. Nano Energy. 2017;**39**:562-570. DOI: 10.1016/j.nanoen.2017.06.018

[42] Pu X, Li LX, Liu MM, Jiang CY, Du CH, Zhao ZF, et al. Wearable self-charging power textile based on flexible yarn supercapacitors and fabric nanogenerators. Advanced Materials. 2016;**28**:98. DOI: 10.1002/adma.201504403

[43] Chen J, Guo HY, Pu XJ, Wang X, Xi Y, Hu CG. Traditional weaving craft for one-piece self-charging power textile for wearable electronics. Nano Energy. 2018;**50**:536-543. DOI: 10.1016/j.nanoen.2018.06.009

[44] Zhong QZ, Zhong JW, Hu B, Hu QY, Zhou J, Wang ZL. A paper-based nanogenerator as a power source and active sensor. Energy & Environmental Science. 2013;**6**:1779-1784. DOI: 10.1039/c3ee40592c

[45] Yang PK, Lin ZH, Pradel KC, Lin L, Li XH, Wen XN, et al. Paper-based origami triboelectric nanogenerators and self-powered pressure sensors. ACS Nano. 2015;**9**:901-907. DOI: 10.1021/nn506631t

[46] Xia KQ, Zhu ZY, Zhang HZ, Du CL, Xu ZW, Wang RJ. Painting a high-output triboelectric nanogenerator on paper for harvesting energy from human body motion. Nano Energy. 2018;**50**:571-580. DOI: 10.1016/j.nanoen.2018.06.019

[47] Yang Y, Zhang HL, Lin ZH, Zhou YS, Jing QS, Su YJ, et al. Human skin based triboelectric nanogenerators for harvesting biomechanical energy and as self-powered active tactile sensor system. ACS Nano. 2013;**7**:9213-9222. DOI: 10.1021/nn403838y

[48] Dai YT, Fu YM, Zeng H, Xing LL, Zhang Y, Zhan Y, et al. A self-powered brain-linked vision electronic-skin based on triboelectric-photodetecing pixel-addressable matrix for visual-image recognition and behavior intervention. Advanced Functional Materials. 2018;**28**:9. DOI: 10.1002/adfm.201800275

[49] Chen HT, Song Y, Guo H, Miao LM, Chen XX, Su ZM, et al. Hybrid porous micro structured finger skin inspired self-powered electronic skin system for pressure sensing and sliding detection. Nano Energy. 2018;**51**:496-503. DOI: 10.1016/j.nanoen.2018.07.001

[50] Khan U, Kim TH, Ryu H, Seung W, Kim SW. Graphene tribotronics for electronic skin and touch screen applications. Advanced Materials. 2017;**29**:8. DOI: 10.1002/adma.201603544

[51] Lai YC, Deng J, Liu R, Hsiao YC, Zhang SL, Peng W, et al. Actively perceiving and responsive soft robots enabled by self-powered, highly extensible, and highly sensitive triboelectric proximity- and pressure-sensing skins. Advanced Materials. 2018;**30**:12. DOI: 10.1002/adma.201801114

[52] Wen Z, Yang YQ, Sun N, Li GF, Liu YN, Chen C, et al. A wrinkled PEDOT:PSS film based stretchable and transparent triboelectric nanogenerator for wearable energy harvesters and active motion sensors. Advanced Functional Materials. 2018;**28**:8. DOI: 10.1002/adfm.201803684

[53] Wang XF, Yin YJ, Yi F, Dai KR, Niu SM, Han YZ, et al. Bioinspired stretchable triboelectric nanogenerator as energy-harvesting skin for self-powered electronics. Nano Energy. 2017;**39**:429-436. DOI: 10.1016/j.nanoen.2017.07.022

[54] Dong K, Wu Z, Deng J, Wang AC, Zou H, Chen C, et al. A stretchable yarn embedded triboelectric nanogenerator as electronic skin for biomechanical energy harvesting and multifunctional pressure sensing. Advanced Materials. 2018;**30**:1804944. DOI: 10.1002/adma.201804944

[55] Wang XD, Zhang YF, Zhang XJ, Huo ZH, Li XY, Que ML, et al. A highly stretchable transparent self-powered triboelectric tactile sensor with metallized nanofibers for wearable electronics. Advanced Materials. 2018;**30**:8. DOI: 10.1002/adma.201706738

[56] Deng JN, Kuang X, Liu RY, Ding WB, Wang AC, Lai YC, et al. Vitrimer elastomer-based jigsaw puzzle-like healable triboelectric nanogenerator

for self-powered wearable electronics. Advanced Materials. 2018;**30**:10. DOI: 10.1002/adma.201705918

[57] Parida K, Kumar V, Wang JX, Bhavanasi V, Bendi R, Lee PS. Highly transparent, stretchable, and self-healing ionic-skin triboelectric nanogenerators for energy harvesting and touch applications. Advanced Materials. 2017;**29**:8. DOI: 10.1002/adma.201702181

[58] Wang SH, Xie YN, Niu SM, Lin L, Wang ZL. Freestanding triboelectric-layer-based nanogenerators for harvesting energy from a moving object or human motion in contact and non-contact modes. Advanced Materials. 2014;**26**:2818-2824. DOI: 10.1002/adma.201305303

[59] Xie YN, Wang SH, Niu SM, Lin L, Jing QS, Yang J, et al. Grating-structured freestanding triboelectric-layer nanogenerator for harvesting mechanical energy at 85% total conversion efficiency. Advanced Materials. 2014;**26**:6599-6607. DOI: 10.1002/adma.201402428

[60] Chen J, Zhu G, Yang WQ, Jing QS, Bai P, Yang Y, et al. Harmonic-resonator-based triboelectric nanogenerator as a sustainable power source and a self-powered active vibration sensor. Advanced Materials. 2013;**25**: 6094-6099. DOI: 10.1002/adma.201302397

[61] Yang J, Chen J, Yang Y, Zhang HL, Yang WQ, Bai P, et al. Broadband vibrational energy harvesting based on a triboelectric nanogenerator. Advanced Energy Materials. 2014;**4**:9. DOI: 10.1002/aenm.201301322

[62] Hu YF, Yang J, Jing QS, Niu SM, Wu WZ, Wang ZL. Triboelectric nanogenerator built on suspended 3D spiral structure as vibration and positioning sensor and wave energy harvester. ACS Nano. 2013;**7**:10424-10432. DOI: 10.1021/nn405209u

[63] Zhang HL, Yang Y, Su YJ, Chen J, Adams K, Lee S, et al. Triboelectric nanogenerator for harvesting vibration energy in full space and as self-powered acceleration sensor. Advanced Functional Materials. 2014;**24**: 1401-1407. DOI: 10.1002/adfm.201302453

[64] Wen XN, Yang WQ, Jing QS, Wang ZL. Harvesting broadband kinetic impact energy from mechanical triggering/vibration and water waves. ACS Nano. 2014;**8**:7405-7412. DOI: 10.1021/nn502618f

[65] Wang XF, Niu SM, Yi F, Yin YJ, Hao CL, Dai K, et al. Harvesting ambient vibration energy over a wide frequency range for self-powered electronics. ACS Nano. 2017;**11**:1728-1735. DOI: 10.1021/acsnano.6b07633

[66] Yang WQ, Chen J, Jing QS, Yang J, Wen XN, Su YJ, et al. 3D stack integrated triboelectric nanogenerator for harvesting vibration energy. Advanced Functional Materials. 2014;**24**:4090-4096. DOI: 10.1002/adfm.201304211

[67] Tang W, Jiang T, Fan FR, Yu AF, Zhang C, Cao X, et al. Liquid-metal electrode for high-performance triboelectric nanogenerator at an instantaneous energy conversion efficiency of 70.6%. Advanced Functional Materials. 2015;**25**:3718-3725. DOI: 10.1002/adfm.201501331

[68] Wang J, Li SM, Yi F, Zi YL, Lin J, Wang XF, et al. Sustainably powering wearable electronics solely by biomechanical energy. Nature Communications. 2016;**7**:8. DOI: 10.1038/ncomms12744

[69] Xu MY, Wang PH, Wang YC, Zhang SL, Wang AC, Zhang CL, et al. A soft

and robust spring based triboelectric nanogenerator for harvesting arbitrary directional vibration energy and self-powered vibration sensing. Advanced Energy Materials. 2018;**8**(9). DOI: 10.1002/aenm.201702432

[70] Yang XY, Chan S, Wang LY, Daoud WA. Water tank triboelectric nanogenerator for efficient harvesting of water wave energy over a broad frequency range. Nano Energy. 2018;**44**:388-398. DOI: 10.1016/j.nanoen.2017.12.025

[71] Pan L, Wang JY, Wang PH, Gao RJ, Wang YC, Zhang XW, et al. Liquid-FEP-based U-tube triboelectric nanogenerator for harvesting water-wave energy. Nano Research. 2018;**11**:4062-4073. DOI: 10.1007/s12274-018-1989-9

[72] Li XY, Tao J, Zhu J, Pan CF. A nanowire based triboelectric nanogenerator for harvesting water wave energy and its applications. APL Materials. 2017;**5**:6. DOI: 10.1063/1.4977216

[73] Lin ZH, Cheng G, Lee S, Pradel KC, Wang ZL. Harvesting water drop energy by a sequential contact-electrification and electrostatic-induction process. Advanced Materials. 2014;**26**:4690. DOI: 10.1002/adma.201400373

[74] Jeon SB, Kim D, Seol ML, Park SJ, Choi YK. 3-Dimensional broadband energy harvester based on internal hydrodynamic oscillation with a package structure. Nano Energy. 2015;**17**:82-90. DOI: 10.1016/j.nanoen.2015.08.002

[75] Choi D, Lee S, Park SM, Cho H, Hwang W, Kim DS. Energy harvesting model of moving water inside a tubular system and its application of a stick-type compact triboelectric nanogenerator. Nano Research. 2015;**8**:2481-2491. DOI: 10.1007/s12274-015-0756-4

[76] Zhu G, Su YJ, Bai P, Chen J, Jing QS, Yang WQ, et al. Harvesting water wave energy by asymmetric screening of electrostatic charges on a nanostructured hydrophobic thin-film surface. ACS Nano. 2014;**8**:6031-6037. DOI: 10.1021/nn5012732

[77] Lee S, Chung J, Kim DY, Jung JY, Lee SH, Lee S. Cylindrical water triboelectric nanogenerator via controlling geometrical shape of anodized aluminum for enhanced electrostatic induction. ACS Applied Materials & Interfaces. 2016;**8**:25014-25018. DOI: 10.1021/acsami.6b08828

[78] Zhao XJ, Zhu G, Fan YJ, Li HY, Wang ZL. Triboelectric charging at the nanostructured solid/liquid interface for area-scalable wave energy conversion and its use in corrosion protection. ACS Nano. 2015;**9**:7671-7677. DOI: 10.1021/acsnano.5b03093

[79] Zhao XJ, Kuang SY, Wang ZL, Zhu G. Highly adaptive solid-liquid interfacing triboelectric nanogenerator for harvesting diverse water wave energy. ACS Nano. 2018;**12**:4280-4285. DOI: 10.1021/acsnano.7b08716

[80] Lin ZH, Cheng G, Li XH, Yang PK, Wen XN, Wang ZL. A multi-layered interdigitative-electrodes-based triboelectric nanogenerator for harvesting hydropower. Nano Energy. 2015;**15**:256-265. DOI: 10.1016/j.nanoen.2015.04.037

[81] Xie YN, Wang SH, Niu SM, Lin L, Jing QS, Su YJ, et al. Multi-layered disk triboelectric nanogenerator for harvesting hydropower. Nano Energy. 2014;**6**:129-136. DOI: 10.1016/j.nanoen.2014.03.015

[82] Kim DY, Kim HS, Kong DS, Choi M, Kim HB, Lee JH, et al. Floating buoy-based triboelectric nanogenerator for an effective vibrational energy harvesting from irregular and random water waves in

wild sea. Nano Energy. 2018;**45**:247-254. DOI: 10.1016/j.nanoen.2017.12.052

[83] Xiong JQ, Lin MF, Wang JX, Gaw SL, Parida K, Lee PS. Wearable all-fabric-based triboelectric generator for water energy harvesting. Advanced Energy Materials. 2017;7:10. DOI: 10.1002/aenm.201701243

[84] Su YJ, Wen XN, Zhu G, Yang J, Chen J, Bai P, et al. Hybrid triboelectric nanogenerator for harvesting water wave energy and as a self-powered distress signal emitter. Nano Energy. 2014;**9**:186-195. DOI: 10.1016/j.nanoen.2014.07.006

[85] Lin ZH, Cheng G, Wu WZ, Pradel KC, Wang ZL. Dual-mode triboelectric nanogenerator for harvesting water energy and as a self-powered ethanol nanosensor. ACS Nano. 2014;**8**: 6440-6448. DOI: 10.1021/nn501983s

[86] Cheng G, Lin ZH, Du ZL, Wang ZL. Simultaneously harvesting electrostatic and mechanical energies from flowing water by a hybridized triboelectric nanogenerator. ACS Nano. 2014;**8**:1932-1939. DOI: 10.1021/nn406565k

[87] Nguyen V, Yang RS. Effect of humidity and pressure on the triboelectric nanogenerator. Nano Energy. 2013;**2**:604-608. DOI: 10.1016/j.nanoen.2013.07.012

[88] Jiang T, Zhang LM, Chen XY, Han CB, Tang W, Zhang C, et al. Structural optimization of triboelectric nanogenerator for harvesting water wave energy. ACS Nano. 2015;**9**:12562-12572. DOI: 10.1021/acsnano.5b06372

[89] Ahmed A, Saadatnia Z, Hassan I, Zi YL, Xi Y, He X, et al. Self-powered wireless sensor node enabled by a duck-shaped triboelectric nanogenerator for harvesting water wave energy. Advanced Energy Materials. 2017;7:10. DOI: 10.1002/aenm.201601705

[90] Jing QS, Zhu G, Bai P, Xie YN, Chen J, Han RPS, et al. Case-encapsulated triboelectric nanogenerator for harvesting energy from reciprocating sliding motion. ACS Nano. 2014;**8**: 3836-3842. DOI: 10.1021/nn500694y

[91] Xi Y, Wang J, Zi YL, Li XG, Han CB, Cao X, et al. High efficient harvesting of underwater ultrasonic wave energy by triboelectric nanogenerator. Nano Energy. 2017;**38**:101-108. DOI: 10.1016/j.nanoen.2017.04.053

[92] Yang Y, Zhang HL, Liu RY, Wen XN, Hou TC, Wang ZL. Fully enclosed triboelectric nanogenerators for applications in water and harsh environments. Advanced Energy Materials. 2013;**3**:1563-1568. DOI: 10.1002/aenm.201300376

[93] Wang XF, Niu SM, Yin YJ, Yi F, You Z, Wang ZL. Triboelectric nanogenerator based on fully enclosed rolling spherical structure for harvesting low-frequency water wave energy. Advanced Energy Materials. 2015;**5**:9. DOI: 10.1002/aenm.201501467

[94] Xu L, Jiang T, Lin P, Shao JJ, He C, Zhong W, et al. Coupled triboelectric nanogenerator networks for efficient water wave energy harvesting. ACS Nano. 2018;**12**:1849-1858. DOI: 10.1021/acsnano.7b08674

[95] Wang ZL. New wave power. Nature. 2017;**542**:159-160

[96] Wu CS, Liu RY, Wang J, Zi YL, Lin L, Wang ZL. A spring-based resonance coupling for hugely enhancing the performance of triboelectric nanogenerators for harvesting low-frequency vibration energy. Nano Energy. 2017;**32**:287-293. DOI: 10.1016/j.nanoen.2016.12.061

[97] Xiao TX, Liang X, Jiang T, Xu L, Shao JJ, Nie JH, et al. Spherical triboelectric nanogenerators based on spring-assisted multilayered structure

for efficient water wave energy harvesting. Advanced Functional Materials. 2018;**28**:8. DOI: 10.1002/adfm.201802634

[98] Chen B, Yang Y, Wang ZL. Scavenging wind energy by triboelectric nanogenerators. Advanced Energy Materials. 2018;**8**:13. DOI: 10.1002/aenm.201702649

[99] Seol ML, Woo JH, Jeon SB, Kim D, Park SJ, Hur J, et al. Vertically stacked thin triboelectric nanogenerator for wind energy harvesting. Nano Energy. 2015;**14**:201-208. DOI: 10.1016/j.nanoen.2014.11.016

[100] Phan H, Shin DM, Jeon SH, Kang TY, Han P, Kim GH, et al. Aerodynamic and aeroelastic flutters driven triboelectric nanogenerators for harvesting broadband airflow energy. Nano Energy. 2017;**33**:476-484. DOI: 10.1016/j.nanoen.2017.02.005

[101] Xie YN, Wang SH, Lin L, Jing QS, Lin ZH, Niu SM, et al. Rotary triboelectric nanogenerator based on a hybridized mechanism for harvesting wind energy. ACS Nano. 2013;**7**:7119-7125. DOI: 10.1021/nn402477h

[102] Li SM, Wang SH, Zi YL, Wen Z, Lin L, Zhang G, et al. Largely improving the robustness and lifetime of triboelectric nanogenerators through automatic transition between contact and noncontact working states. ACS Nano. 2015;**9**:7479-7487. DOI: 10.1021/acsnano.5b02575

[103] Yang Y, Zhu G, Zhang HL, Chen J, Zhong XD, Lin ZH, et al. Triboelectric nanogenerator for harvesting wind energy and as self-powered wind vector sensor system. ACS Nano. 2013;**7**:9461-9468. DOI: 10.1021/nn4043157

[104] Wang SH, Mu XJ, Wang X, Gu AY, Wang ZL, Yang Y. Elasto-aerodynamics-driven triboelectric nanogenerator for scavenging air-flow energy. ACS Nano. 2015;**9**:9554-9563. DOI: 10.1021/acsnano.5b04396

[105] Zhao ZF, Pu X, Du CH, Li LX, Jiang CY, Hu WG, et al. Freestanding flag-type triboelectric nanogenerator for harvesting high-altitude wind energy from arbitrary directions. ACS Nano. 2016;**10**:1780-1787. DOI: 10.1021/acsnano.5b07157

[106] Bae J, Lee J, Kim S, Ha J, Lee BS, Park Y, et al. Flutter-driven triboelectrification for harvesting wind energy. Nature Communications. 2014;**5**:9. DOI: 10.1038/ncomms5929

[107] Chen SW, Gao CZ, Tang W, Zhu HR, Han Y, Jiang QW, et al. Self-powered cleaning of air pollution by wind driven triboelectric nanogenerator. Nano Energy. 2015;**14**:217-225. DOI: 10.1016/j.nanoen.2014.12.013

[108] Kim D, Tcho IW, Choi YK. Triboelectric nanogenerator based on rolling motion of beads for harvesting wind energy as active wind speed sensor. Nano Energy. 2018;**52**:256-263. DOI: 10.1016/j.nanoen.2018.07.046

[109] Wen Z, Chen J, Yeh MH, Guo HY, Li ZL, Fan X, et al. Blow-driven triboelectric nanogenerator as an active alcohol breath analyzer. Nano Energy. 2015;**16**:38-46. DOI: 10.1016/j.nanoen.2015.06.006

[110] Wu YC, Zhong XD, Wang X, Yang Y, Wang ZL. Hybrid energy cell for simultaneously harvesting wind, solar, and chemical energies. Nano Research. 2014;**7**:1631-1639. DOI: 10.1007/s12274-014-0523-y

[111] Qian JG, Jing XJ. Wind-driven hybridized triboelectric-electromagnetic nanogenerator and solar cell as a sustainable power unit for self-powered natural disaster monitoring sensor networks. Nano Energy. 2018;**52**:78-87. DOI: 10.1016/j.nanoen.2018.07.035

[112] Wen Z, Yeh MH, Guo HY, Wang J, Zi YL, Xu WD, et al. Self-powered textile for wearable electronics by hybridizing fiber-shaped nanogenerators, solar cells, and supercapacitors. Science Advances. 2016;**2**:8. DOI: 10.1126/sciadv.1600097

[113] Cao R, Zhou T, Wang B, Yin YY, Yuan ZQ, Li CJ, et al. Rotating-sleeve triboelectric-electromagnetic hybrid nanogenerator for high efficiency of harvesting mechanical energy. ACS Nano. 2017;**11**:8370-8378. DOI: 10.1021/acsnano.7b03683

[114] Maharjan P, Toyabur RM, Park JY. A human locomotion inspired hybrid nanogenerator for wrist-wearable electronic device and sensor applications. Nano Energy. 2018;**46**:383-395. DOI: 10.1016/j.nanoen.2018.02.033

[115] Quan T, Wang X, Wang ZL, Yang Y. Hybridized electromagnetic-triboelectric nanogenerator for a self-powered electronic watch. ACS Nano. 2015;**9**:12301-12310. DOI: 10.1021/acsnano.5b05598

[116] Zhang KW, Wang X, Yang Y, Wang ZL. Hybridized electromagnetic-triboelectric nanogenerator for scavenging biomechanical energy for sustainably powering wearable electronics. ACS Nano. 2015;**9**: 3521-3529. DOI: 10.1021/nn507455f

[117] Hu YF, Yang J, Niu SM, Wu WZ, Wang ZL. Hybridizing triboelectrification and electromagnetic induction effects for high-efficient mechanical energy harvesting. ACS Nano. 2014;**8**:7442-7450. DOI: 10.1021/nn502684f

[118] Zhong XD, Yang Y, Wang X, Wang ZL. Rotating-disk-based hybridized electromagnetic-triboelectric nanogenerator for scavenging biomechanical energy as a mobile power source. Nano Energy. 2015;**13**:771-780. DOI: 10.1016/j.nanoen.2015.03.012

[119] Guo HY, Wen Z, Zi YL, Yeh MH, Wang J, Zhu LP, et al. A water-proof triboelectric-electromagnetic hybrid generator for energy harvesting in harsh environments. Advanced Energy Materials. 2016;**6**:7. DOI: 10.1002/aenm.201501593

[120] Lee S, Bae SH, Lin L, Ahn S, Park C, Kim SW, et al. Flexible hybrid cell for simultaneously harvesting thermal and mechanical energies. Nano Energy. 2013;**2**:817-825. DOI: 10.1016/j.nanoen.2013.02.004

[121] Guo YB, Zhang XS, Wang Y, Gong W, Zhang QH, Wang HZ, et al. All-fiber hybrid piezoelectric-enhanced triboelectric nanogenerator for wearable gesture monitoring. Nano Energy. 2018;**48**:152-160. DOI: 10.1016/j.nanoen.2018.03.033

Chapter 5

RF Energy Harvesting System and Circuits for Charging of Wireless Devices Using Spectrum Sensing

Naser Ahmadi Moghaddam and Alireza Maleki

Abstract

Recently, lots of works have been done on the optimal power management of wireless devices. This leads to the main idea of ambient energy harvesting. Among various energy harvesting approaches, one is to use radio waves existing in the ambient environment for battery charging, called RF energy harvesting. In this chapter, in order to improve the RF energy harvesting performance, we utilize spectrum sensing to allow the wireless devices to select the frequency band with maximum power that exceeds a predefined threshold to charge the device (this power threshold can be determined according to battery type and its required charging power) and the device can use this power for battery charging. Also, a novel voltage multiplier circuit is proposed. By means of simulations and experimental tests, it can be seen that after detection of our desired 1 mW RF signal, system output power is about 532μ W and 450μ W in simulation and practical situations respectively.

Keywords: energy harvesting, voltage multiplier, OFDM, spectrum sensing

1. Introduction

Recently, with the rapid growth of wireless communication systems, researchers have studied various challenges about improvement of these systems from many aspects such as performance, error optimization, hardware design and implementation and etc. with introducing the wireless sensor networks, Internet of Things (IoT) and robotics, one of the main challenges appeared is the energy consumption of these systems and how to provide reliable and low cost power supply to feed these systems as long as possible with high durability. That is the main reason for all of the researches conducted on energy harvesting. Various methods and approaches are presented to tackle the issue e.g. improvement of batteries structure and their capacity, piezoelectric materials and movement of human body part to produce the required power mostly for wearable devices, or thermal and magnetic energy harvesting approaches.

Mainly, there are two energy sources: mechanical and magnetic waves [1]. For energy harvesting purposes, as in [2–5], natural sources like solar energy wind, vibrations and movement of human body parts or magnetic waves can be exploited. Here in this chapter, our focus is on a specific kind of electromagnetic source, Radio Frequency (RF) signal, which is produced by the oscillation of photons in a certain frequency and used for transmitting data and information in communication systems.

Examples of these transceivers in today's world are Frequency Modulation (FM) radio, Analog TV (ATV), Digital TV (DTV), mobile and cellular networks and Wi-Fi. To have a more clear understanding of the issue and seeing RF signals as a energy source, in **Figure 1**, DTV and cellular signal spectrums for Tokyo City and Yokohama City are indicated [6]. As it can be seen in this figure, for some certain frequencies, the measured power is about 0 and −20 dB.

By saying RF energy harvesting, we mean that we capture the energy from the RF signal existing in the ambient and transform this power into DC power and using it for supplying battery. Passive ambient RF energy harvesting is exactly defined as this procedure [7]. In this case, sources can be FM radio, Wi-Fi, DTV or military communication transmitters [8] and the amount of energy harvested from these sources is in the order of 1 to 10 $\frac{\mu W}{cm^2}$ [9]. Also there is another type of RF energy harvesting, i.e. RF energy harvesting from a dedicated source. In this scenario, the amount of harvested energy is higher comparing to passive ambient case and is in the order of 50 $\frac{\mu W}{cm^2}$ [9]. One example for this category is RFID chips [10].

1.1 Preprocessing in energy harvesting system

Spectrum is a scarce source. In wireless communication systems, efforts have been made to use frequency spectrum with policies and priorities in order to maximize the spectrum efficiency. The main idea here is to allocate empty spectrum holes over time, frequency and space to secondary users while the interference with primary user is minimum. Several approaches are proposed for spectrum sensing, such as energy detection [11–13], matched filter [11, 12, 14], cyclostationary detection [15, 16], spectrum sensing based on covariance matrix [17, 18] and wavelet based spectrum sensing [19]. By studying energy detection, it can be understood that this approach is based on detecting the signal power such a way that secondary users detect the signal power received from primary users. Then they compare it to some predefined threshold level and then decide whether they can use the primary frequency band or not. Well, here is the novel preprocessing idea which we are exploited in this chapter:

"In RF energy harvesting, we use RF signal power and convert it to DC power for charging batteries. On the other hand, energy detection algorithms give us the

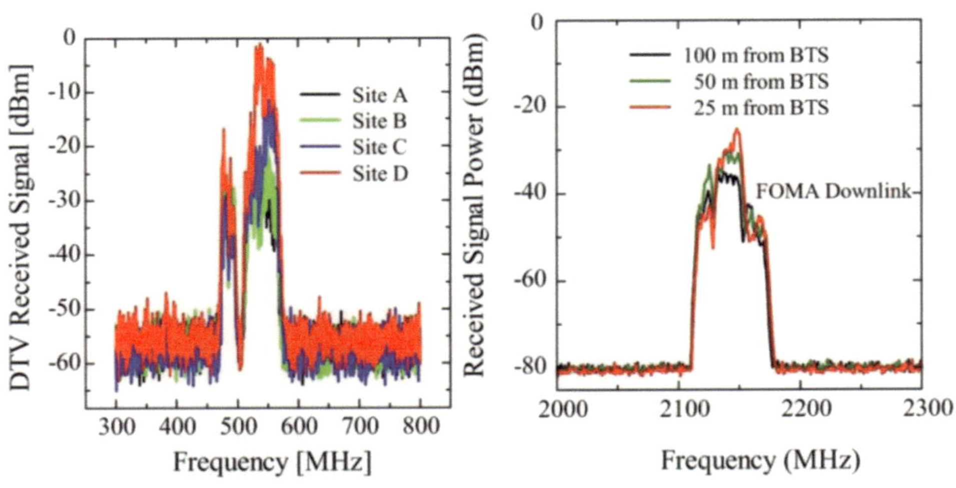

Figure 1.
DTV signal spectrum measured in Tokyo City (left side graph) and Cellular signal spectrum measured in Yokohama City (right side graph) [6].

ability to detect the ambient signal power and compare it to a predefined threshold level. So, we do the same here as well. But with the goal of detecting a strong RF signal. In other words, by exploiting this preprocessing, we compare our received RF signal power and if it is greater than a threshold, then we will decide that frequency contains our required power for energy harvesting purposes and switch our circuit to that frequency. In the simulation process, this threshold is set to 0 dBm which is a reasonable and practical assumption based on **Figure 1**."

2. Proposed method

Our proposed system is indicated in **Figure 2**. As it is indicated in this figure, by preprocessing stage, the frequency containing the high amount of energy is selected. After that, this signal is selected as the input of matching circuit and rectified. Then a DC-DC converter circuit is used to level up the DC signal and finally it is fed to the battery for charging.

2.1 Battery model

For simulation stage and performance evaluation of our proposed system, we must be able to model the battery that we intend to charge. There are various battery models with different structures and complexities. Electrochemical models [20–22] are usually used for battery physical design, performance and power generation optimization. Mathematical models [23–26], are much more effective. Random events for predicting battery systematic behaviors like battery life time and efficiency are discussed using mathematical equations. Electrical models [27–33] are placed somewhere between mathematical and chemical models in terms of accuracy and utilize the combination of voltage sources, resistors and capacitors.

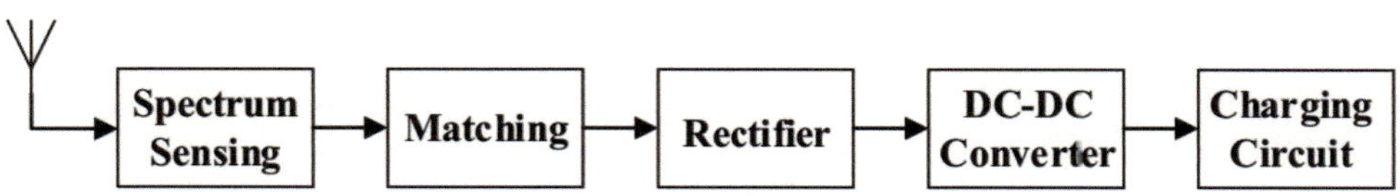

Figure 2.
Schematic of proposed system.

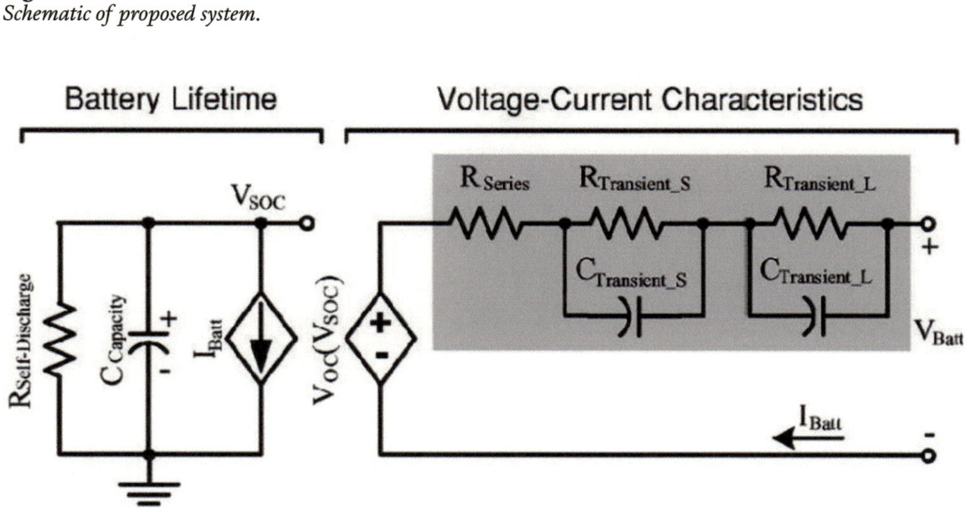

Figure 3.
Battery model [34].

For the goal of this chapter, an accurate and effective battery model is proposed based on battery model proposed in [34]. This model provides an easy extraction procedure, gives run time, static and transient responses and also contains all of the electrodynamic characteristics of the batteries. **Figure 3** shows this proposed model.

3. Results

Simulation results and fabrication of our system is presented in this section. Spectrum sensing, circuits and antenna are simulated using MATLAB, Advance Design System (ADS) and Computer Simulation Technology (CST) software respectively. We consider the frequency band 500 MHz to 1 GHz for maximum power RF signal extraction. At the end, our fabricated charger circuit is tested in laboratory environment.

3.1 Spectrum sensing simulations

We consider M random base stations in the aforementioned frequency band. Using OFDM transmitter, the output signal is generated and by exploiting energy detection, our desired RF signal for energy harvesting circuit is obtained. Parameters of simulation are given in **Table 1**. Note that for digital to analog converter, a13 th order Butterworth filter is used (normalized cut off frequency equals to $\frac{1}{20}$).

RC pulse shaping time response is as follows [35].

$$S_{RC}(t) = sinc\left(\frac{t}{T_s}\right) \times \frac{\cos\left(\frac{\pi\alpha t}{T_s}\right)}{1 - \frac{4\alpha^2 t^2}{T_s^2}} \tag{1}$$

where α and T_s are roll-off factor (takes values from 0 to 1) and symbol repetition rate respectively.

As it can be seen in **Figure 4**, there are four signals available in the spectrum and in order to find the RF signal with maximum power, the area under each signal should be calculated i.e. its power. After applying energy detection we find that in this case, maximum power RF signal happened at 915 MHz. Therefore, the filter is set to select this signal out of the spectrum.

Parameters	Value (OFDM)
Number of sub-carriers	2048
Occupied of sub-carriers	1024
Sampling frequency (MHz)	20
Number of oversampling	4
Pulse shaper	Raised cosine
Number of random frequency	4
IFFT length	4096
Bandwidth (MHz)	10

Table 1.
Stimulation parameters.

After finding the frequency in which our desired RF signal exists, spectrum should be fed to a filter with the central frequency of 915 MHz and the bandwidth of 10 MHz. In **Figure 5**, filter characteristics in terms of frequency response is given. Note that this filter has the ability to be tuned to select the maximum frequency each time. Also, this filter should reject the rest of frequency band otherwise we face some challenges such as power loss and circuit design complexity.

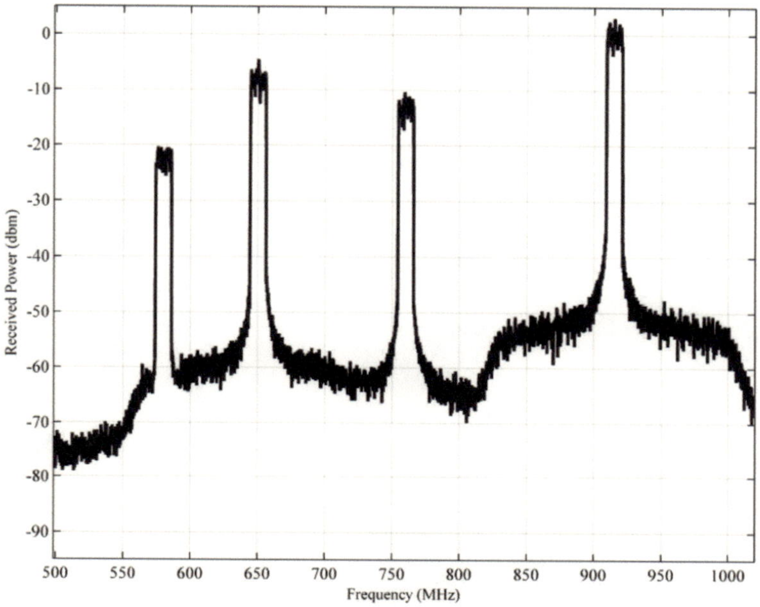

Figure 4.
Frequency response of OFDM transmitter in our frequency band with QPSK modulation in receiver end.

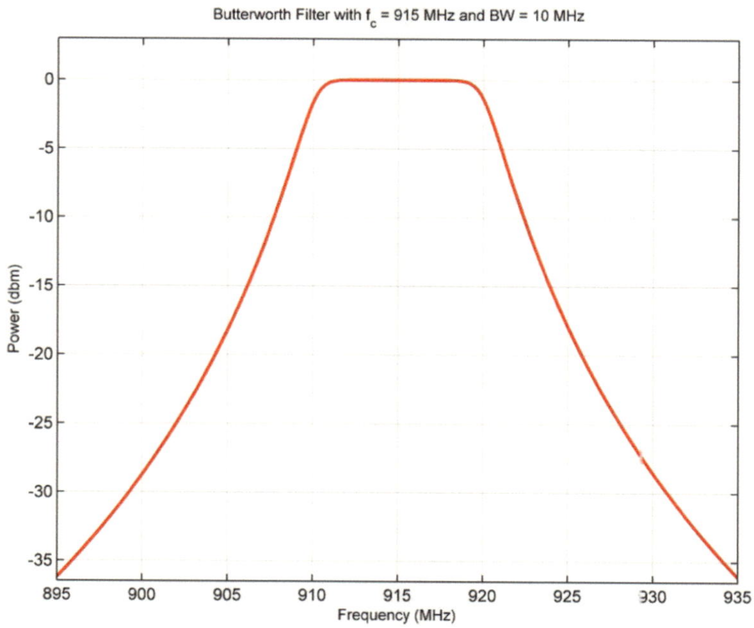

Figure 5.
Frequency response of Butterworth filter with central frequency of 915 MHz and bandwidth of 10 MHz.

Figure 6 shows power for filter, spectrum and the output of filter (Antenna input signal). We use OFDM transmitter as mentioned earlier with QPSK modulation.

3.2 Circuit design and simulation

In previous section, RF signal with maximum power is identified and extracted from spectrum. Now, this signal represents the input of our charging circuit. We assume a1 mW signal with the frequency of 915 MHz as our input to execute our

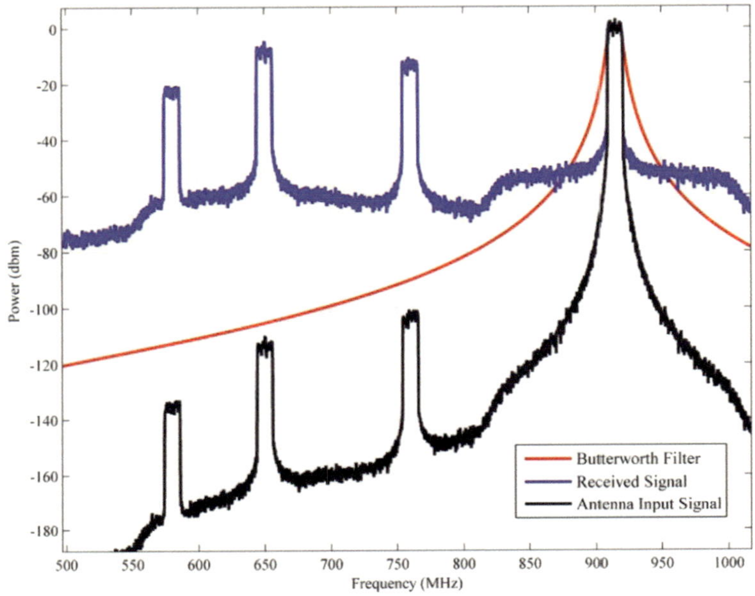

Figure 6.
Frequency response of power spectrum, filter and filter output for OFDM transmitter with QPSK modulation.

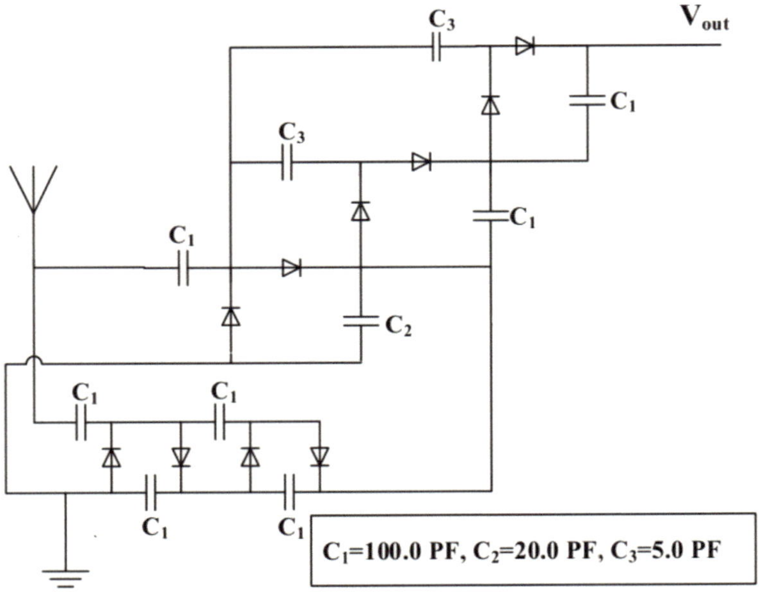

Figure 7.
Proposed voltage multiplier circuit.

simulation in ADS. First, we need to increase the DC level of our signal using a DC voltage multiplier circuit as it is shown in **Figure 7**. Note that we have a 4-stage and a 6-stage voltage multiplier in our proposed circuit. The connection between these two multipliers results in increasing of the output voltage.

In **Figure 8**, Efficiency of the rectifier circuit is shown versus different values of P_{in}. As it can be seen, when the input power is 0 dbm, highest efficiency is obtained at the frequency of 915 MHz.

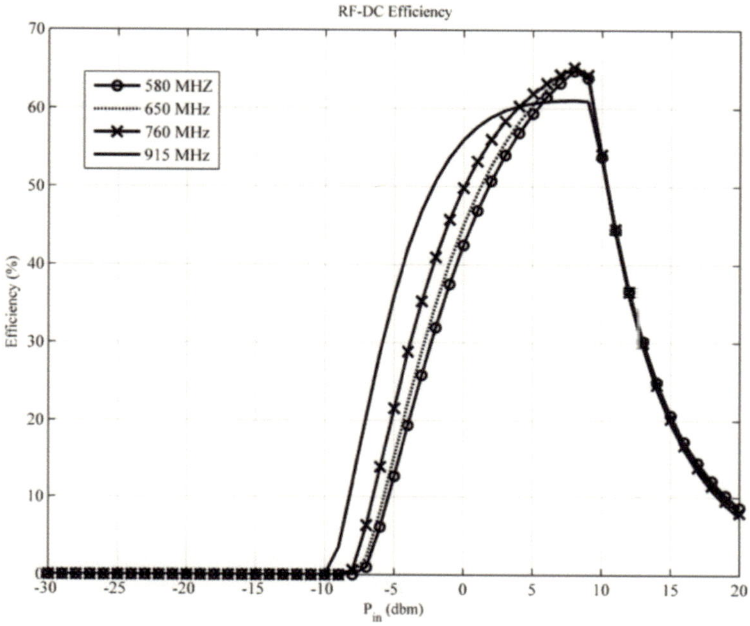

Figure 8.
Rectifier circuit efficiency.

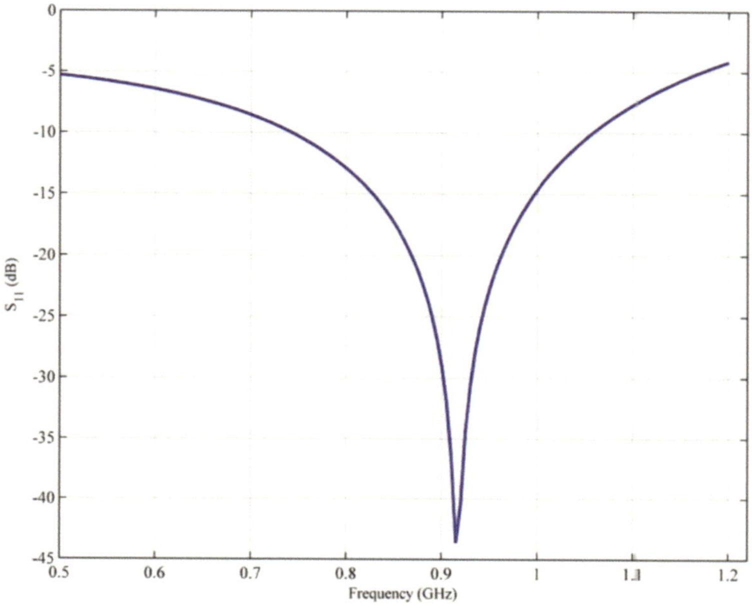

Figure 9.
Rectifier circuit return loss (s 11) at 0 dbm input RF power.

Figure 9 indicates the return loss (s_{11}) for different frequencies and minimum return loss is obtained at 915 MHz frequency at a circuit input power of 0 dbm. This is resulted from our designed matching circuit and shows its desirable performance.

In **Figure 10**, rectifier circuit output voltage and current are indicated with the maximum at 915 MHz.

For frequencies of 580, 650, 760 and 915 MHz, the outputs of voltage multiplier circuit are shown in **Figure 11**. As it can be seen, we obtain a 8.8 V DC voltage for our input signal.

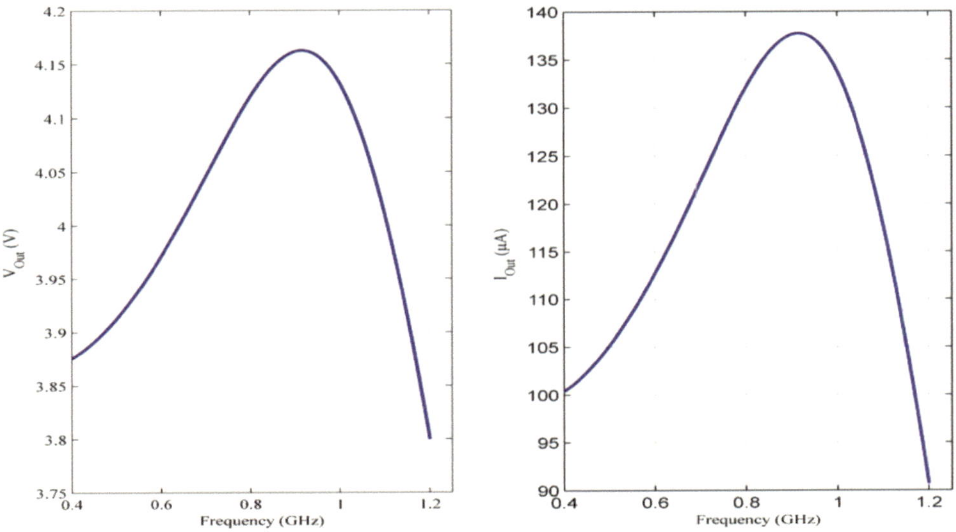

Figure 10.
Rectifier circuit output voltage and current.

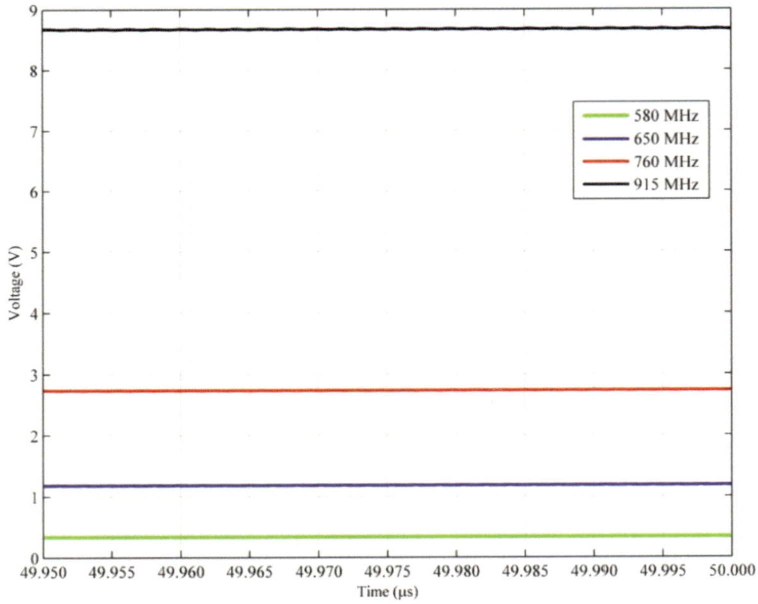

Figure 11.
Voltage multiplier circuit output.

Figure 12 shows the higher efficiency of our proposed circuit comparing to three other methods. That is because of exploiting a 10-stage voltage multiplier (a 4-stage connected to a 6-stage).

Complete charger circuit is proposed in **Figure 13**. Output currents for aforementioned four frequencies are given in **Figure 14**. Also, output voltages are

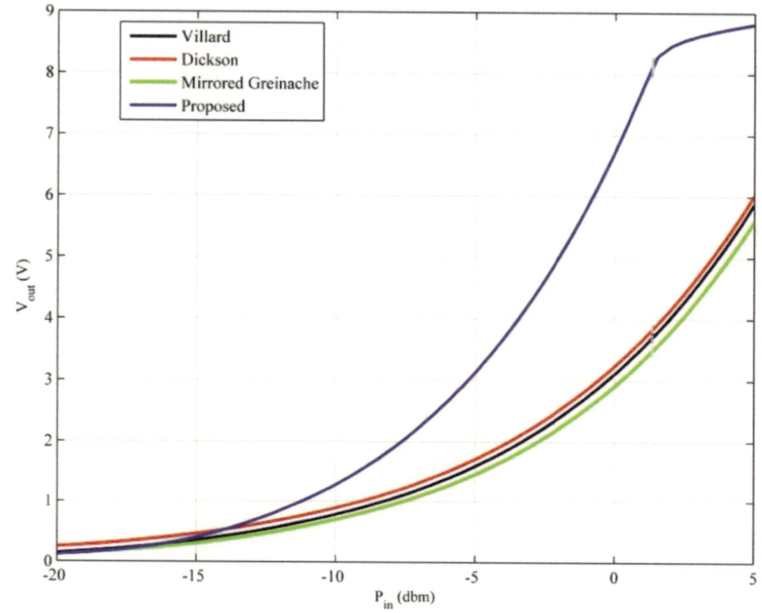

Figure 12.
Output voltage versus input power.

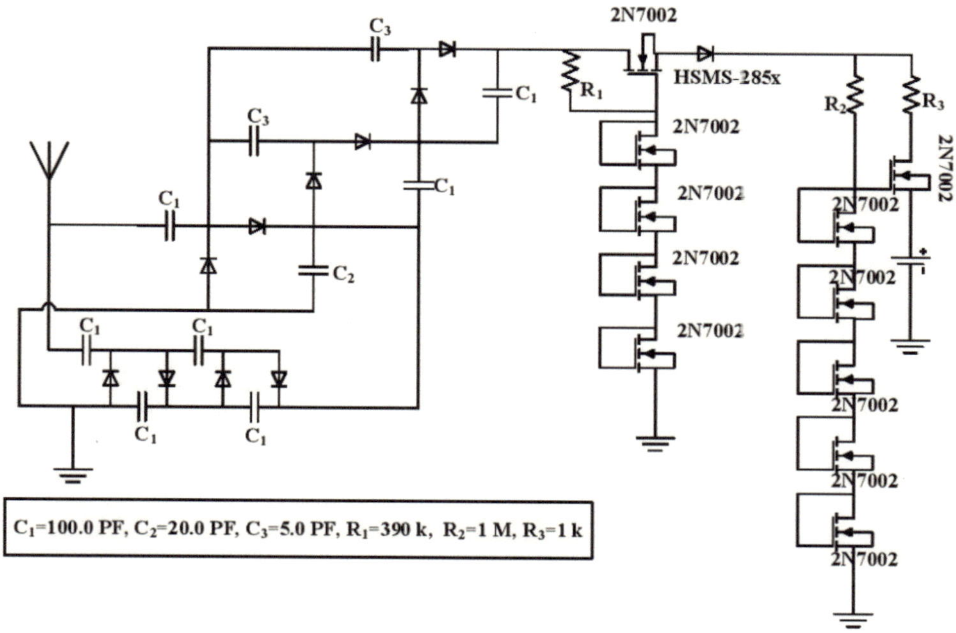

Figure 13.
Charger circuit.

indicated in **Figure 15**. So knowing that power equals to current times voltage, we obtain 532μ W output power for 915 MHz signal. It should be noted that voltage drop in 915 MHz (8.8 V to 3.6 V) is because that by connecting battery to charger circuit, battery charging process starts and in this process power must be constant, so by increasing the current drawn by the battery, output voltage drops.

3.3 Antenna simulation

Receiver antenna is a broadband type with linearly polarized radiation. Considering that our bandwidth is 500 MHz, according to previous experiences, we

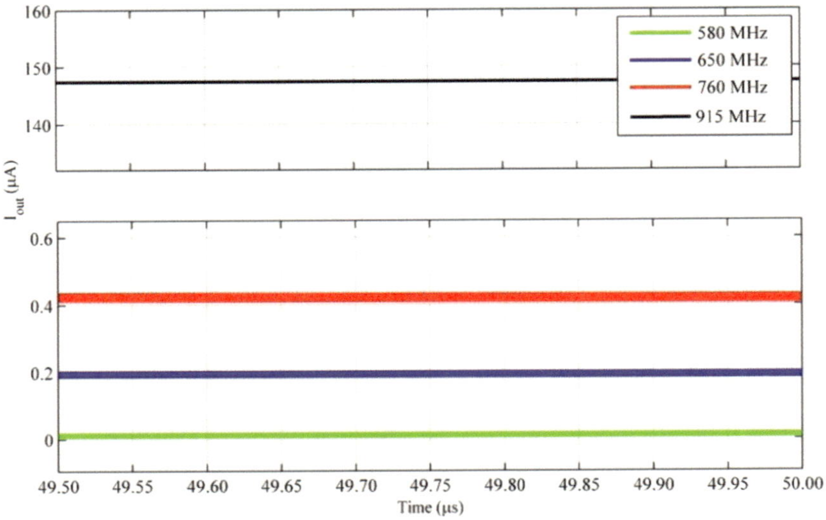

Figure 14.
Output currents in 580, 650, 760 and 915 MHz.

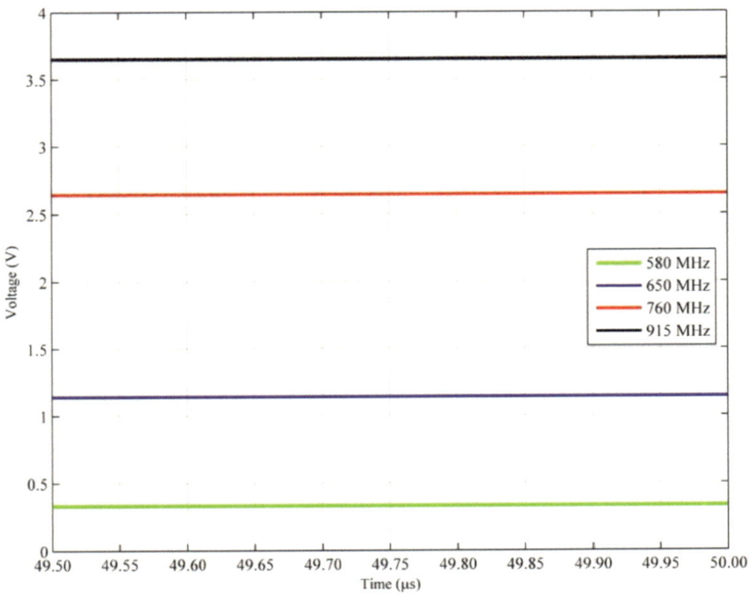

Figure 15.
Output voltages in 580, 650, 760 and 915 MHz.

require an Ultra Wide Band (UWB) antenna (**Figure 16**). By means of simulation and measurement, reflection coefficients of our designed antenna are indicated in **Figure 17**. Note that because of the size of this antenna, we did not use it obtain fabrication results. A challenge may be rise to decrease the size of UWB antenna and improve our designed system performance in the future.

3.4 Charger circuit fabrication

Test results for our fabricated charger circuit are indicated in **Figure 18**. In terms of experimental results, we obtain 6.89 V of output voltage at the frequency of 915 MHz (which obtained 8.8 V in simulation results), where our input signal is a 0 dbm generated by R&S®SMB100A signal generator. For a 10 kΩ resistive load, load voltage is 2.12 V and we can calculate the output power as 450μ W.

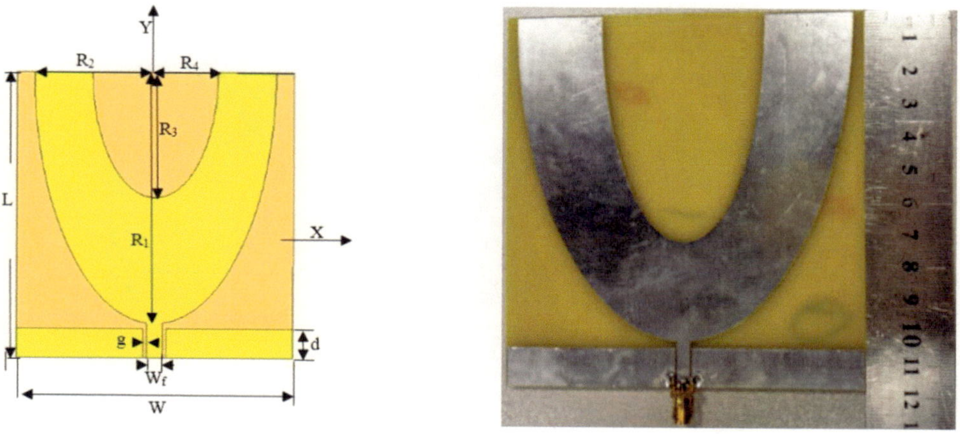

(a) (b)

Figure 16.
Designed U-shape UWB antenna.

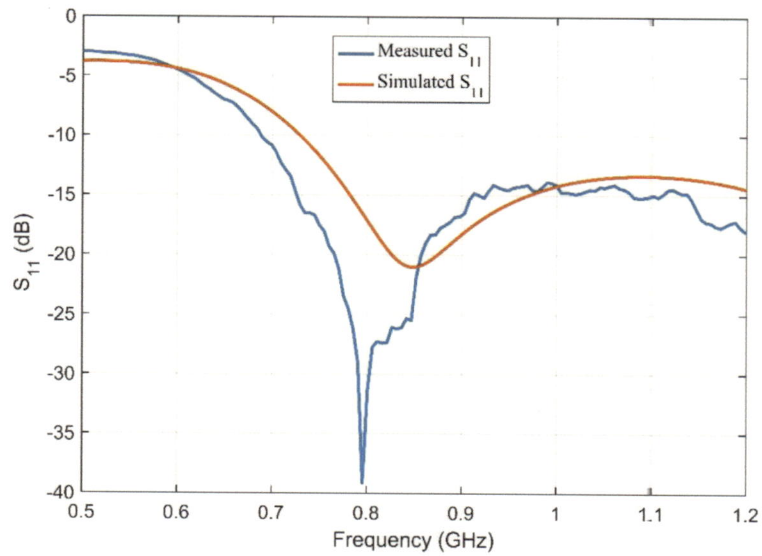

Figure 17.
Measured and simulated reflection coefficients of UWB antenna.

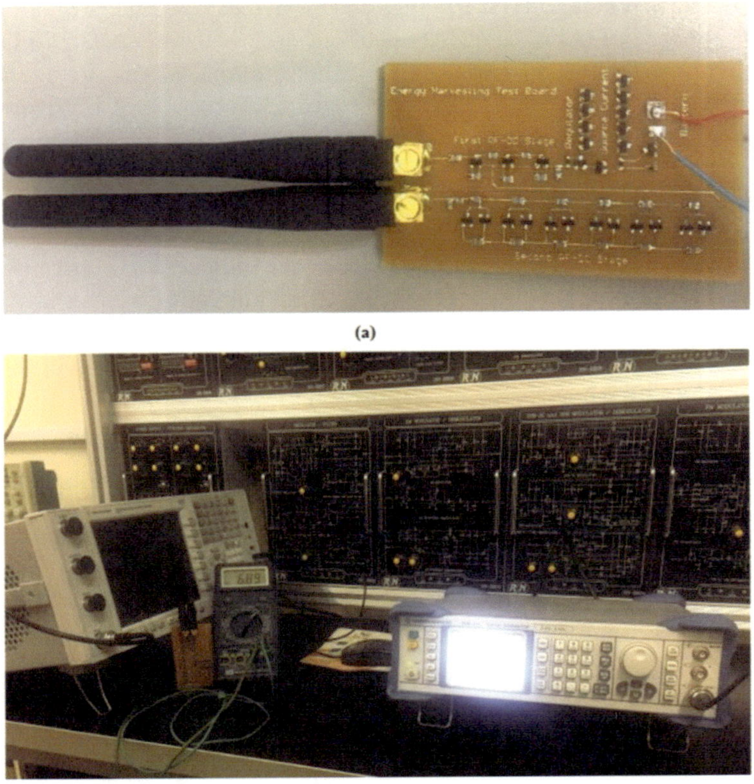

(a)

(b)

Figure 18.
(a) Fabricated charger circuit (b) Circuit test.

4. Conclusion

A novel approach for RF energy harvesting is presented in this chapter. Unlike previous works, we exploit spectrum sensing as a pre-processing procedure to find the frequency in which the maximum power more than a predefined threshold exists. After that, using a filter, this signal with maximum power is fed to a charger circuit. In simulation, a DC output voltage of 8.8 V is obtained using input RF power of 0 dbm. In practice and as our experimental results, our fabricated charger circuit test led to 450μ W output power for battery charging.

Author details

Naser Ahmadi Moghaddam[1*] and Alireza Maleki[2]

1 Tarbiat Modares University, Tehran, Iran

2 K.N. Toosi University of Technology, Tehran, Iran

*Address all correspondence to: n.moghaddam1990@gmail.com

References

[1] Soyata T, Copeland L, Heinzelaman W. RF energy harvesting for embedded systems: A survey of tradeoffs and methodology. IEEE Circuits and Systems Magazine. 2016;**16**(1):22-57

[2] Zhang JY, Cao Z, Wang Q, Kuwano H. Microstructure and piezoelectric properties of AlN thin films grown on stainless steel for the application of vibration energy harvesting. IET Micro & Nano Letters. 2012;**7**(12):1170-1172

[3] Lafarge B, Curea O, Hacala A, Camblong H. Analysis, design & simulation of an electromechanical energy harvesting system using a linear movement. In: International Conference on Green Energy; 25-27 March 2014

[4] Ali QI. Event driven duty cycling: An efficient power management scheme for a solar-energy harvested road side unit. IET Electrical Systems in Transportation. 2016;**6**(3):222-235

[5] Azevedo JAR, Santos FES. Energy harvesting from wind and water for autonomous wireless sensor nodes. IET Circuits, Devices and Systems. 2012;**6**(6):413-420

[6] Tan YK. Sustainable Energy Harvesting Technologies—Past, Present and Future. Rijeka, Croatia: InTech; 2011. Ch. 10

[7] Paradiso J, Starner T. Energy scavenging for mobile and wireless electronics. IEEE Pervasive Computer. 2005;**4**(1):18-27

[8] Intel Corp. Intel ambient RF energy harvesting demonstration [Online]. Available from: http://www.rfwirelesssensors.com/2009/01/intel-ambientrf-energy-harvesting-demonstration/

[9] Honan G, Gekakis N, Hassanalieragh M, Nadeau A, Sharma G, Soyata T. Energy harvesting and buffering for cyber physical systems: A review. In: Cyber Physical Systems—A Computational Perspective, ch. 7. 2015. United States: CRC; pp. 191-217. ISBN. 978-1-4822-5975-9

[10] Impinj Inc. SPEEDWAY IPJ-R1000 Reader/Interrogator [Online]. Available from: http://www.impinj.com/

[11] Sahai A, Cabric D. Spectrum sensing: Fundamental limits and practical challenges. In: IEEE International Symposium on New Frontiers in Dynamic Spectrum Access Networks, Baltimore, November; 2005

[12] Kay SM. Fundamentals of Statistical Signal Processing: Detection Theory. 2nd ed. NJ, USA: Prentice Hall, Upper Saddle Rive; 1998

[13] Urkowitz H. Energy detection of unknown deterministic signals. Proceedings of the IEEE. 1967;**55**(4):523-531

[14] Gao W, Daut DG, Chen HS. Signature based spectrum sensing algorithms for IEEE802.22 WRAN. In: IEEE International Conference on Communications, Glasgow; 24-28 June 2007; pp. 6487-6492

[15] Gardner WA. Exploitation of spectral redundancy in cyclostationary signals. IEEE Signal Processing Magazine. 1991;**8**(2):14-36

[16] Han N, Shon S, Chung JH, Kim JM. Spectral correlation based signal detection method for spectrum sensing in IEEE802.22 WRAN systems. In: The 8th International Conference on Advanced Communication Technology, Phoenix Park; Vol. 3. 2006. pp. 1765-1770

[17] Zeng Y, Liang YC. Covariance based signal detections for cognitive radio. In: 2nd IEEE International Symposium on New Frontiers in Dynamic Spectrum Access Networks, Dublin; 17-20 April 2007; pp. 202-207

[18] Zeng Y, Liang YC. Maximum-minimum eigenvalue detection for cognitive radio. In: 18th International Symposium on Personal, Indoor and Mobile Radio Communications, Athens; 3-7 September 2007; pp. 1-5

[19] Wornell GW. Emerging applications of multirate signal processing and wavelets in digital communications. Proceedings of the IEEE. 1996;**84**(4):586-603

[20] Song L, Evans JW. Electrochemical-thermal model of lithium polymer batteries. Journal of the Electrochemical Society. 2000;**147**:2086-2095

[21] Gomadam PM, Weidner JW, Dougal RA, White RE. Mathematical modeling of lithium-ion and nickel battery systems. Journal of Power Sources. 2002;**110**(2):267-224

[22] Newman J, Thomas KE, Hafezi H, Wheeler DR. Modeling of lithium-ion batteries. Journal of Power Sources. 2003;**119-121**:838-843

[23] Pedram M, Wu Q. Design considerations for battery-powered electronics. In: Proceedings of the 36th Annual ACM/IEEE Design Automation Conference; 1999. pp. 861-866

[24] Rynkiewicz R. Discharge and charge modeling of lead acid batteries. Applied Power Electronics Conference and Exposition. 1999;**2**:707-710

[25] Rakhmatov D, Vrudhula S, Wallach DA. A model for battery lifetime analysis for organizing applications on a pocket computer. IEEE Transanctions on VLSI Systems. 2003;**11**(6):1019-1030

[26] Rong P, Pedram M. An analytical model for predicting the remaining battery capacity of lithium-ion batteries. IEEE Transactions on Very Large Scale Integration (VLSI) Systems. 2006;**14**(5):441-451

[27] Valvo M, Wicks FE, Robertson D, Rudin S. Development and application of an improved equivalent circuit model of a lead acid battery. In: Proceedings of the 31st Intersociety Energy Conversion Engineering Conference. Vol. 2. 1996. pp. 1159-1163

[28] Ceraolo M. New dynamical models of lead-acid batteries. IEEE Transanctions on Power Systems. 2000;**15**(4):1184-1190

[29] Barsali S, Ceraolo M. Dynamical models of lead-acid batteries: Implementation issues. IEEE Transanctions on Energy Conversion. 2002;**17**(1):16-23

[30] Gao L, Liu S, Dougal RA. Dynamic lithium-ion battery model for system simulation. IEEE Transactions on Components and Packaging Technologies. 2002;**25**(3):495-505

[31] Glass MC. Battery electrochemical nonlinear/dynamic SPICE model. In: Proceedings of the 31st Intersociety Energy Conversion Engineering Conference. Vol. 1. 1996. pp. 292-297

[32] Baudry P, Neri M, Gueguen M, Lonchampt G. Electro-thermal modeling of polymer lithium batteries for starting period and pulse power. Journal of Power Sources. 1995;**54**(2):393-396

[33] Hageman SC. Simple pspice models let you simulate common battery types. EDN. 1993;**38**:17-132

[34] Chen M, Rincon-Mora GA. Accurate electrical battery model capable

of predicting runtime and I-V performance. IEEE Transactions on Energy Conversion. 2006;**21**(2)

[35] Nyquist H. Abridgment of certain topics in telegraph transmission theory. Journal of the A.I.I.E. 1928;**47**(3):214-217

Chapter 6

Thousand kW High-Temperature Solar Furnace in Parkent (Uzbekistan) – Energetical Characteristics

Akbarov Rasul

Abstract

This chapter presents a method of calculation of the energetical characteristics of the large solar furnace with a capacity of 1000 kW (LSF) taking into account its real optical parameters. The technical characteristics of the LSF are presented. The possible energy characteristics of the LSF based on numerical calculations are analyzed. The energy characteristics of the total system with different inaccuracies of the reflecting surfaces, energy contributions of certain shelves and groups of heliostats, and the contributions of certain heliostats and shapes of their focal spot are determined. Empirical formulas are proposed to describe the obtained numerical results. The problem of implementing the possible energy modes of the LSF with and/or without the inclusion of certain shelves and groups of heliostats is analyzed. The problem of a day changes in the energy density distribution in the focal spot of the LSF is considered.

Keywords: solar radiation, thermal power, furnace, concentrator, heliostat, focal zone, energy distribution, numerical calculation

1. Introduction

One of the promising directions in the use of solar energy is materials science. As is known, high-temperature heating by solar radiation has certain advantages, e.g., the absence of pollution from synthesized materials, instantaneous heating, the possibility to control the heating and cooling rate, a wide range of solar radiation, etc. At present, high-temperature solar technologies are widely applied in many areas of science and engineering. In this respect, concentrated solar energy is an important component among the available material synthesis methods with a set of specified properties [1–8].

The important characteristics of the technological processes are the capacity, maximum and average energy densities, uniformity of the energy density distribution, focal spot size, character of the energy density distribution and its change in time, duration of the process, start and stop mode rate, etc.

A characteristic drawback of solar concentrators is variability of the characteristics of the focal spot with time. This is related, on the one hand, with the temporary change in the direct solar radiation value, and on the other hand, with the condition of the optical and mechanical elements of solar furnaces, i.e., the adjustment condition of the mirrors, angular inaccuracies of the light-reflecting mirrors, reflection coefficient of the mirrors, condition of the solar sensors, etc.

The Materials Science Institute of the Academy of Sciences of the Republic of Uzbekistan operates a large solar furnace of 1000 kW thermal power (LSF). It is a composite optical and mechanical complex with automatic control systems, comprising the heliostat field (62 heliostats), paraboloidal concentrator (with 1906 m^2 midship section area-projection of the surface of the concentrator on the plane), and technological tower. Highly concentrated solar radiation, with energy density up to 700 W/cm^2, may be nowadays created in the focal region of the LSF, located in the technological tower. A set of specialized testing facilities are developed at the LSF, having not analogues in terms of a number of parameters either in home or foreign practice. Note that the abbreviation BSF (Big Solar Furnace) is often used instead of the LSF.

The LSF was put into operation as an experimental-industrial one, in the summer of 1987 [9, 10].

During Soviet times this unique solar furnace was the leading facility of military-industrial complex for testing of different materials and equipment to the action of concentrated solar radiation and for development of advanced ceramic materials for high-tech industry.

In connection with the change of research direction (from military to peaceful), the fields of technology of engineering ceramics, fireproof materials of wide range of applications, particularly for metallurgy, oil and gas complex, power engineering, machine building, and chemical industries became the major directions of use of LSF.

Scientists in Uzbekistan have achieved significant results in high-temperature solar technologies. More than 150 compositions of various oxide materials having unique properties and serving as the basis for functional, structural, and high-refractory ceramics have been developed and synthesized at the LSF, and their thermophysical and other characteristics have been studied. The LSF is a unique instrument for field studies of high-temperature processes, i.e., on the synthesis and heat treatment of materials and study of their properties.

Note that such a furnace was previously commissioned in Odeillo, France. This 1000 kW solar furnace was completed in 1970 (**Figure 1**). The furnace contains 63 orientated mirrors (heliostats), each of 45 m^2 surface, with 180 single mirror

Figure 1.
General view of the France furnace.

panes (facets). The movement of this installation is guided by electronic controls operated by reflected rays. The accuracy of the control is 1 min of arc, but due to dispersion on the flat glass, 5 min of arc are obtained on the reflected rays. A solar beam of constant energy is directly horizontally southward to a paraboloid reflector of 2000 m^2 intercepted area. This paraboloid contains 9500 single glass panes, bent by mechanical constraint and adjusted to reflect maximum radiation to the focal plane, situated 18 m from the apex of the parabola [3, 4, 11].

Technological capabilities of the furnace are described in the works [2–5].

2. Main technical characteristics of the LSF

The LSF as noted above represents a complex optomechanical aggregate with automatic control systems consisting of a heliostat field and paraboloid concentrator, which form a high-density radiation flux in the focal zone of the concentrator. The furnace is located 50 km from Tashkent, in the Parkent District. The geographical location is 41.32°N, 69.74°E; its altitude above sea level is 1050 m.

The LSF heliostat field is formed by 62 heliostats located on the smooth slope of a mountain (slope 13°) in a checkerboard pattern arranged on 8 terraces. All 62 heliostats of the LSF have a similar structure and dimension. The reflecting surface of the heliostat with dimensions of 7.5 × 6.5 m is flat and composite and consists of 195 facets with dimensions of 0.5 × 0.5 m and a thickness of 6 mm.

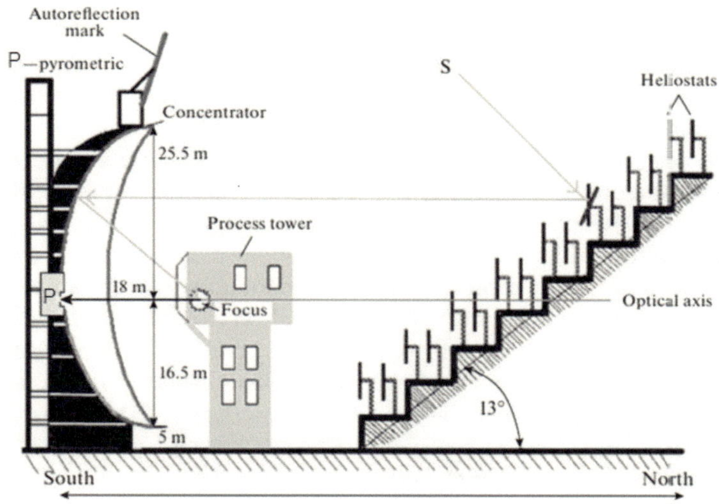

a

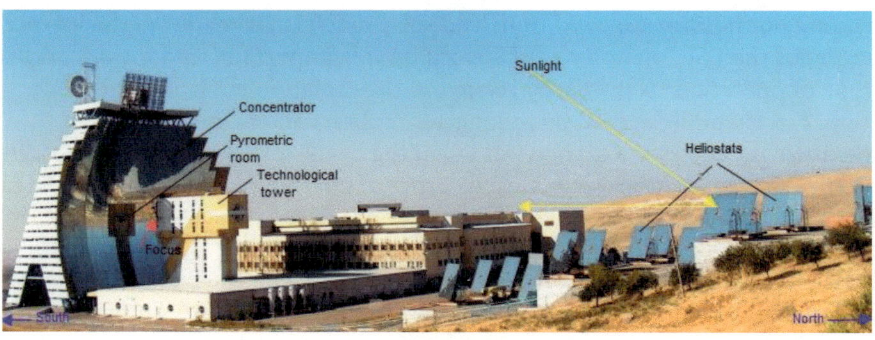

b

Figure 2.
(a) Overview diagram of LSF and (b) general view of the LSF.

The reflecting surface of the concentrator is a rectangular-stepped cutting of a paraboloid of revolution with a focal distance of 18 m. The height of the midsection of the concentrator is 42 m, and the width is 54 m. The total area of the midsection of the reflecting surface is 1906 m^2. The concentrator consists of 214 blocks. 50 rhomb-shaped facets with sides of 447 mm and different apex angles are installed on each block. The thickness of mirrors is 5 mm.

In total, there are 22,790 facets; the total area of the reflecting mirrors (facets) is about 5200 m^2; the heliostat field, process tower, and concentrator occupy approximately 2 ha of ground area. Technical characteristics of the installation are also described in the works [9, 10, 12, 13].

The general pattern of the device is shown in **Figure 2a** and the general view of the installation were shown in **Figure 2b**.

3. Operating experience of the LSF

Some advantages of the LSF should be noted as compared with other solar furnaces, resulting from the specific nature of the optical circuit of the furnace and latest engineering developments used in the furnace's operation: the heliostat automated control system (HACS), computer vision system (CVS), photometers, radiometers, melting furnaces of different designs, modern meteorological station in the territory of the furnace, high-accuracy methods for arranging the furnace's reflecting mirrors, various tests and optical benches, etc [14–21].

Some projects have been implemented in recent years at the entity "Sun" of the Academy of Sciences of the Republic of Uzbekistan, where the LSF is located; the aim of these projects is to improve its operational characteristics and extend its functional capabilities.

One of the most significant finished projects is the commissioning of the modern meteorological station on the territory of the "Sun" entity in 2013 with the financial support of the Asian Development Bank. This station is equipped with high-frequency pyranometers and pyrheliometers to measure solar radiation. The station measures global horizontal radiation and diffuse horizontal and beam normal solar radiation [20]. As is known, beam normal solar radiation is important for the LSF, and depending on its level, it is possible to implement certain high-temperature processes. With the commissioning of the meteorological station, it is possible to define more correctly and promptly the energy characteristics of the furnace using the calibration coefficients. Certain types of operations, such as, preventive maintenance, adjustment and cleaning of the reflecting elements, regulation of the solar sensors of the heliostats, different high-temperature processes requiring certain process modes in the focal zone of the furnace, etc., are performed at the LSF depending on the climatic conditions and solar radiation level. With the intensive operation of the LSF, planning of the work on it is important, and the selection of the type of operations is primarily related to the radiation environment of the location. For example, according to long-term observations, many high-temperature processes are carried out when the beam solar radiation has a value of 600 W/m^2 or larger. At present, various types of operations are planned in advance at the LSF according to the time of day and season based on statistical processing of meteorological data.

Efficient operation of the LSF significantly depends on the technical parameters of the HACS for tracking the Sun due to the specific nature of its optical circuit. At present, the capabilities of the existing HACS developed from outmoded element bases do not meet modern requirements, although it functions. This is why a new modern HACS has been developed at the LSF in recent years; this will make it possible to significantly extend its functional capabilities. The new HACS differs fundamentally from the old one: modern microprocessor circuits are used in the element bases, achievements in modern information technologies have been implemented, reliable operation is ensured, and it has high accuracy and flexibility for the required processing modes [19]. The operational efficiency of the new system has successfully been tested on one heliostat, and efforts on transferring the system to other heliostats are planned (2019-2020). As a result, the LSF will have some new possibilities for practical use.

It should be noted that the third modernization of the control system of the heliostats of the French furnace has also been completed recently [22].

Furthermore, any complex processing device without a system for monitoring and measuring its operating parameters and characteristics cannot function efficiently and reliably. In this respect, the LSF is equipped with modern devices to measure the density of concentrated solar energy and the temperature of the studied materials and devices to control the behavior of the high-temperature processes. They include a high-temperature pyrometer for remote measurement of the temperature of materials (IMPAC IGA 12, T_{max} = 3500°C), a thermal imager (FLIR A655), various radiometers, photometers, digital thermometers, a computer vision system (developments of the Institute of Materials Science - IMS) and etc.

Thus, the LSF, in addition to an environmentally friendly melting furnace, is a unique research tool for high-temperature studies, which has some important advantages compared to other high-temperature solar furnaces:

1. High power of the furnace, up to 1 MW.

2. High levels of solar radiation concentration, ≈ 4500–10,000.

3. Large size of the focal spot of the furnace, ≈ 80–100 cm.

4. Sixty-two heliostats, which makes it possible to control the focal energy distribution.

5. The HACS makes it possible to flexibly control movement of the heliostats.

6. Simulation of 62 aperture-specific concentrators.

7. Control of the behavior of technological processes.

8. The ability to measure and control the optical energy condition of the furnace.

4. Calculation of the optical-energetical characteristics of the LSF: the general approach

In the helio-engineering practice, different techniques for evaluation of energetical characteristics of mirror concentrating systems (MCS) are employed, depending on the purpose, demanded accuracy, etc [23–29]. However, if a big MCS as a solar furnace in Odeillo (France) or Large Solar Furnace near Tashkent (Uzbekistan) is concerned, there are specific features which must be taken into account.

An important feature of a big-scale MCS is the use of multi-heliostat system to create illumination field on its focal zone. Thanks to this feature, there is a possibility of a flexible control upon flux density distribution by individual control of each heliostat operation regime. This is a very important point for many technological processes and special investigations.

Exposure distribution of the whole system is the sum of individual exposure distributions created by particular heliostats. Because of this fact, it is enough to consider the problem of illumination distribution created by a particular heliostat for arbitrary moment and given direction of beams reflected from the heliostat.

We will use the Cartesian coordinate system with axes: X (direction to Zenith), Y (to West) and Z (to South); the center of this system is in the focus of the furnace. Let the direction of beams reflected from the heliostat be characterized by a unit vector $\vec{K}\left(K_x, K_y, K_z\right)$. Then the unit vector of the heliostat's normal is determined by the expression

$$\vec{N_g} = \frac{\vec{S} + \vec{K}}{\sqrt{2\left(1 + \vec{S}\vec{K}\right)}}$$

where $\vec{S}$ is the coordinate of the Sun's unit vector

$$S_x = \sin h_s = \cos\varphi \cos\delta \cos kt + \sin\varphi \sin\delta$$
$$S_y = \cos h_s \sin A_s = \cos\delta \sin kt$$
$$S_z = \cos h_s \cos A_s = \sin\varphi \cos\delta \cos kt - \cos\varphi \sin\delta$$

where δ is the declination of the Sun, k = 15 deg./h, φ is the geographical latitude, and t is time.

The angle coordinates of the heliostat A_n and h_n are determined by expressions

$$N_{gx} = \sin h_n, \quad N_{gy} = \cos h_n \cos A_n, \quad N_{gz} = \cos h_n \sin A_n$$

It is well known that the energy density at arbitrary point A(x_a, y_a, z_a) of the receiver is determined by integration over the last reflecting surface according to the formula [23]:

$$E = \int_{\omega} B d_{\omega_z} = \int_{\omega} B \frac{dS\left(\vec{N_m}\,\overrightarrow{MA}\right)}{MA^4}\left(\overrightarrow{MA} \cdot \vec{N_A}\right) \tag{1}$$

where $\vec{N_m}$, normal to the surface at point M; dS, elementary area around point M; $\vec{N_A}$, normal of the receiving surface; B, energetical brightness in a reflected beam.

A generalized approach for determining B is presented in [23].

Integration boundaries in this formula are determined by the coordinate of the heliostat center position, direction of the reflected beams $\vec{K}$, and asimuthal and local angles A_n and h_n, which determine the heliostat's orientation.

As it is known in large-scale MCS, usually each heliostat illuminates the particular target area on the concentrator at the given moment [13]. It is easy to see that in the case of rectangle heliostat, the projection of its borders onto the concentrator midship section (a projection of concentrator surface into XY plane) will be different ellipses. Determination of the integration boundaries is a rather difficult problem due to complicated shapes. That is why it is desirable to move to the coordinate system connected with the particular heliostat. In this case the integrating boundaries will coincide with the heliostat borders.

Let us determine the Jacobian of this transformation. The origin of the new system **O** is placed at the heliostat's center; axes OX_1 and OY_1 are directed along the height and width correspondingly and the OZ_1 axis along the heliostat's normal. To determine the Jacobian transformation, we apply geometrical definitions. Consider elementary area dS on the surface of the paraboloidal concentrator. For the normal $\vec{N}_m$ at this point, we have

$$\vec{N}_m = \left(\frac{-x}{\sqrt{x^2+y^2+p^2}}, \frac{-y}{\sqrt{x^2+y^2+p^2}}, \frac{p}{\sqrt{x^2+y^2+p^2}}\right)$$

where p is the focal parameter of the paraboloid.

A projection of the elementary surface on dS along the direction of the vector $\vec{K}$ at this point will be equal to

$$dS_K = \frac{dS}{\vec{N}_m \vec{K}}$$

In its turn a projection of this area onto the heliostat surface is determined from the expression

$$dS_g = dx_1 dy_1 = \frac{dS_K}{\vec{N}_g\, \vec{K}} = \frac{dS}{\left(\vec{N}_g\, \vec{K}\right)\left(\vec{N}_m\, \vec{K}\right)}$$

On the other hand, the projection of the elementary area onto the midship section of the concentrator dS_m has the form

$$dS_m = dx\, dy = \frac{dS}{N_{mz}}$$

From these relations we obtain the following expression for the Jacobian transformation:

$$D = \left|\frac{dS_m}{dS_g}\right| = \left|\frac{\left(\vec{N}_g\, \vec{K}\right)\left(\vec{N}_m\, \vec{K}\right)}{N_{mz}}\right| \tag{2}$$

We note that the above presented vector expressions are written taking into account unit normal of the vectors $\vec{N}_g$, $\vec{N}_m$ and $\vec{K}$. Substituting explicit expressions for these vectors into the last formula, finally we get

$$D = \left|\frac{\left(xK_x + yK_y + pK_z\right)(K_x \sin\ h_n + K_y \cos\ h_n \cos\ A_n + K_z \cos\ h_n \sin A_n}{p}\right|$$

For the particular case, when $\vec{K}$ $(0, 0, 1)$ we obtain obvious formula

$$D = |\cos h_n\ \cos A_n|$$

Below we apply the developed approach to calculation of energetical brightness at given point of the receiver taking into account errors of reflecting surfaces, shadowing and blocking effects, etc.

At first the relation between the new and old variables should be determined. Let us consider point G with coordinates $(x_1, y_1, 0)$ on the heliostat's reference frame. Transformation of this point into the other reference frame FXYZ is performed by the following expression:

$$\begin{pmatrix} x_f \\ y_f \\ z_f \end{pmatrix} = T_1(A_n)T_2(-h_n)\begin{pmatrix} x_1 \\ y_1 \\ 0 \end{pmatrix} + \begin{pmatrix} x_0 \\ y_0 \\ z_0 \end{pmatrix}$$

where T_1 and T_2 are anticlockwise rotation matrixes around X and Y axes, respectively [30], and (x_0, y_0, z_0) are the coordinates of the heliostat's center in the FXYZ frame. Now we have to find the intersection point of the vector $\vec{K}$ with the surface of the paraboloid, i.e., solve the following set of equations

$$\left.\begin{array}{l} \dfrac{x - x_f}{K_x} = \dfrac{y - y_f}{K_y} = \dfrac{z - z_f}{K_z} \\ x^2 + y^2 - p^2 + 2pz = 0 \end{array}\right\}$$

from which we obtain

$$x = x_f + tK_x, \quad y = y_f + tK_y, \quad z = z_f + tK_z,$$

where

$$t = \frac{-b + \sqrt{b^2 - ac}}{a}, a = K_x^2 + K_y^2, b = x_f K_x + y_f K_y + pK_z, c = x_f^2 + y_f^2 - 2p_{z_f}$$

Exposure B at the given target point A is determined by the inverse ray tracing method when a ray is directed from this point to the current point M on the concentrator, then a normal to the reflecting surface at this point is determined, and the ray reflected in the heliostat's direction is found. After that the crossing point of this ray with the given heliostat's reflective surface and the corresponding normal at this point is determined. Finally, the direction of the last reflected ray is found, and depending on the fact, whether this ray hits solar disk (determine the angle φ) or not ($B = 0$), the energetical brightness is determined. Distribution of the brightness over the solar disk is taken into consideration according to the Jose formula [23]:

$$B(\varphi) = 1.23\frac{E_0}{\pi\varphi_0} \cdot \frac{1 + 1.5641\sqrt{1 - \frac{\sin^2(\varphi)}{\sin^2(\varphi_0)}}}{2.5641},$$

where φ_0 is the apparent angular size of the solar disk (16 min) and E_0 is the solar radiation.

During the ray tracing procedure, shadowing and blocking conditions of the given heliostat by other heliostats, technological tower, and other installations are taken into account.

The real normal is given with reference to the vector $\vec{N}_m$. For this purpose an auxiliary Cartesian reference frame $X^1Y^1Z^1$ is introduced with the origin at point M and Z axis, directed along the vector $\vec{N}_m$, i.e.,

$$\vec{L_3} = (L_{3x}, L_{3y}, L_{3z}) = \vec{N_m}\ (N_{mx}, N_{my}, N_{mz}),$$

$$\text{where } \left(\vec{L_1}, \vec{L_2}, \vec{L_3}\right) \text{ are directing unit vectors}$$

The direction of unit vectors may be chosen in different ways, depending on the character of deviation of the real vector from the ideal one. Let the unit vectors lay on the meridional and sagittal sections. In this case the unit vectors $\vec{L_1}$ and $\vec{L_2}$ can be determined via the vector product of previously determined vectors:

$$\vec{L_2}\ (L_{2x}, l_{2y}, L_{2z}) = \vec{L_3} \times (0, 0, -1) = \frac{(-L_{3y}, L_{3x}, 0)}{\sqrt{L_{3x}^2 + L_{3y}^2}}$$

Now, the unit vector $\vec{L_1}$ can be determined using $\vec{L_2}$ and $\vec{L_3}$:

$$\vec{L_1} = \vec{L_2} \times \vec{L_3} = \frac{\left[L_{3x}L_{3z}, L_{3z}L_{3y}, -\left(L_{3x}^2 + L_{3y}^2\right)\right]}{\sqrt{L_{3x}^2 + L_{3y}^2}}$$

The angle between the ideal direction $\vec{N}_m$ and real normal γ, which characterizes angle errors of the surface, is unambiguously determined by two angles μ and ν, which are actually its projections onto two mutually orthogonal planes, for example, meridional and sagittal planes [23]. It is easy to see that the angle γ is determined via μ and ν, according to the following relation:

$$tg^2\gamma = tg^2\mu + tg^2\nu \tag{3}$$

Or if smallness of angles μ and ν is taken into account

$$\gamma \cong \sqrt{\mu^2 + \nu^2} \tag{4}$$

Now the following expression may be written for components of the real normal:

$$N_{xr} = \cos\gamma\ tg\mu, \quad N_{yr} = \cos\gamma\ tg\nu, \quad N_{zr} = \cos\gamma$$

Error density distribution of BSF reflecting surfaces for angles μ and ν can be determinate by experiments. Here we supposed that these errors are to be distributed according to the Gaussian function. For a set of random values μ and ν, individually having Gaussian distribution with zero mean, a two-dimensional probability distribution in the absence of correlations has the form [31].

$$W(\mu, \nu) = \frac{1}{2\pi\sigma_\mu\sigma_\nu} \exp\left[-\frac{1}{2}\left(\frac{\mu^2}{\sigma_\mu^2} + \frac{\nu^2}{\sigma_\nu^2}\right)\right] \tag{5}$$

where σ_μ^2, σ_ν^2 are dispersions of corresponding random values.

Generation of random numbers with a given distribution function is the special issue. To obtain pseudorandom numbers with Gaussian distribution function, we used (after thorough testing) the following expression:

$$\xi = a + \sigma\left[\sin(2\pi\eta_1)\sqrt{-2\ln(\eta_2)}\right]$$

where η_1 and η_2 are random numbers uniformly distributed over the interval [0:1]. Note that high-level programming languages have such built-in functions.

Note that in the case of Rayleigh distribution, i.e., when $\sigma_\mu = \sigma_\nu = \sigma$, an explicit expression for the distribution density (Eq. (5)) may be derived. Omitting simple transformations we obtain the following result:

$$W(\gamma) = \frac{\gamma}{\sigma^2}\exp\left(-\frac{\gamma^2}{2\sigma^2}\right), \quad \gamma > 0 \tag{6}$$

as this takes place

$$\sigma_\gamma = \sqrt{D\gamma} = \sigma\sqrt{2}$$

So, the distribution for deviations of the angle γ does not obey the Gaussian law. Experimental results obtained for mirrors of the BSF also indicate the non-Gaussian character of error distributions.

Now the vector $\vec{N}_r$ must be transformed to the basic reference frame:

$$N_{mx} = L_{1x}N_{xr} + L_{2x}N_{yr} + L_{3x}N_{zr}, \quad N_{my} = L_{1y}N_{xr} + L_{2y}N_{yr} + L_{3y}N_{zr},$$

$$N_{mz} = L_{1z}N_{xr} + L_{2z}N_{yr} + L_{3z}N_{zr}$$

Real normals for other reflecting parts may be determined in the same way. In the case when total errors have distribution law distinct from Gaussian, direction of the real normal can be determined using the error density distribution function [31].

It should be emphasized that in view of the character of the integrand function, chosen approach to solution of the problem, specific features of the algorithm, and possibilities of modern computers, it is desirable to calculate the integral (Eq.(1)) by the statistical method (Monte Carlo procedure) [31]. Moreover, in the problem under consideration, simultaneous determination of exposure values is possible in all given target points (not only at one point) using one set of random numbers is possible. In this way calculation steps are significantly reduced.

Corresponding software based on the above described technique has been developed in Delphi programming language. Using these programs case calculations of the BSF energetical characteristics are performed.

5. Characteristic features of the energy modes of the LSF

5.1 Preliminary notes and some assumptions

Analysis of the characteristics of the processing modes of material synthesis and heat treatment, i.e., isochronic maps of the energy density distribution in the focal zone of the LSF, shows that they are very different. When referring to the energy characteristics of the furnace, it is necessary to consider many factors, which are in a certain temporary state. This is why, when specifically referring to the energy state of the furnace, it is necessary to provide these factors with the corresponding information. Despite this, some characteristic features of the energy parameters of the LSF can be refined, and it is necessary to analyze a large amount of information on the energy density distribution in the focal zone from certain heliostats, shelves, and groups of heliostats for the various system conditions in order to determine them.

The aim of this paragraph is a detailed analysis of the LSF energy characteristics based on numerical calculations. The peculiarities of the methods for calculating the LSF energy characteristics and their implementation for specific problems are given in [25, 28, 29]. Some characteristic features of the LSF energy characteristics are given in these works but with no detailed theoretical or design analysis.

Each heliostat illuminates a certain area of the concentrator in the normal operation mode of the device. A scaled circuit of the concentrator midsection with the block circuits (solid line) and relevant heliostat zones (dotted line) is shown in **Figure 3**. The numbers of the heliostats are given in the left angle of their zone. The upper contour line of the building roof adjacent to the process tower, which insignificantly blocks the light flux from the heliostats, is also shown in **Figure 3**. As is clear from **Figure 3**, the heliostats 55 and 62 are most inefficient (less than 50% of the heliostat area is used).

The LSF energy spot is formed from the energy contributions (irradiance/energy density) of certain heliostats. The energy contributions of the heliostats depend on the place of their location, reflection coefficient, mirror inaccuracy, adjustment state, etc.

The authors developed a program to calculate the energy characteristics taking into account the real influencing factors in order to study the peculiarities of the LSF energy characteristics [25]. The program uses Monte Carlo method to calculate

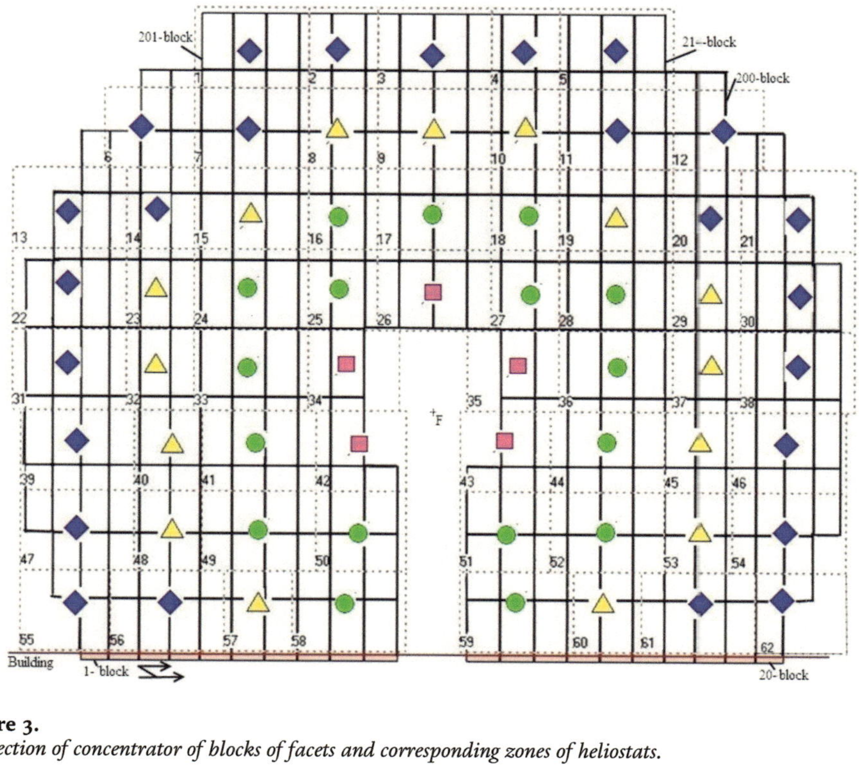

Figure 3.
Midsection of concentrator of blocks of facets and corresponding zones of heliostats.

the integral. Comparative analysis of the calculation results shows that the calculation accuracy is 2–4%. The intensity of the solar beam is determined by backtracking of the beam path. The program interface is shown in **Figure 4**, along with the layout of the heliostats and numbers of the heliostats of certain shelves.

The input parameters of the program are the beam solar radiation E_0 [W/cm^2]; R_{sc} and R_{sg} are the reflection coefficients of the mirrors of the concentrators and heliostats, the root-mean-square mirror inaccuracy of the heliostats σ_g and concentrator σ_c, the relative boundaries of integration of the heliostats, the number of points in the focal plane, the number of playouts in the Monte Carlo method, etc. The energy density distribution was determined for a flat receiver located in the focal plane of the concentrator in all calculations.

In the solar energy problems, the reflecting properties of the real mirror are characterized by the root-mean-square angular inaccuracy σ defined as a root-mean-square value of deviations of the real normal from the ideal (γ). It is assumed in many calculations that the meridional (μ) and sagittal (ν) components of the angle of deviation from the normal ($\gamma^2 \approx \mu^2 + \nu^2$) obey a Gaussian law for integral paraboloid concentrators [23]. The sagittal component of deviation v is often neglected. However, in the case of the LSF, due to the large sizes of the device, the curvature of some concentrator facets (with dimensions of 45 × 45 cm in various sections), especially the peripheral facets, is very similar and small; i.e., the facets are almost isotropic reflectors. The reflecting surfaces of the heliostats also consist of flat reflectors. For this reason, it is assumed in calculations that the probability distribution of the deviation of the mirror normal γ obeys a Gaussian law with dispersion σ. In other words, if the direction of the real normal relative to the ideal is expressed by the formula

$$l_x = \sin\gamma\cos\varphi, \quad l_y = \sin\gamma\sin\varphi, \quad l_z = \cos\gamma,$$

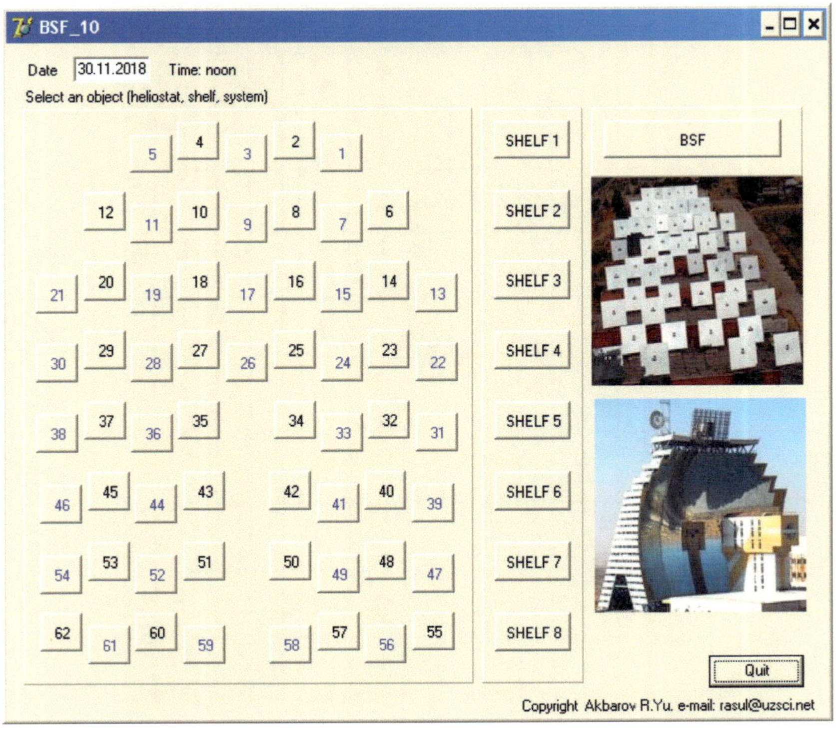

Figure 4.
Program interface.

γ varies according to a Gaussian law with dispersion σ, and φ varies uniformly within 0–2π.

This program was used to carry out some sequences of the numerical calculation to determine the irradiance distribution (energy density) of certain heliostats, shelves of heliostats, groups of heliostats, and the entire LSF with different inaccuracies of the reflecting surfaces of the heliostats and concentrator (1–10 angular minutes).

5.2 Analysis of the results of numerical calculations

5.2.1 Analysis of the energy characteristics of the whole system

The scope of calculations is quite large, which is why some peculiarities of the LSF energy characteristics are revealed from analysis of data, which are given below.

Figure 5a shows the dependence of the focal irradiance E_F on the mirror inaccuracy of the heliostats σ_g and concentrator σ_c (E_0 = 700 W/m^2, $R_{sc} = R_{sg}$ = 0.6).

The mirror inaccuracies of the heliostats σ_g are measured on the horizontal axis. Some individual lines correspond to various inaccuracies of the concentrator mirrors σ_c. It is easy to see that the change in focal irradiance is expectedly more sensitive to mirror inaccuracies of the concentrator than the heliostats. However, beginning with 2.5–3.0 angular minutes, the influence of the mirror inaccuracies of the heliostats and concentrator on the character of the irradiance distribution does not significantly differ and is approximately similar. In practice, we have the mirror inaccuracies within 3–10 angular minutes. This is why, for decreasing (by an order of magnitude) calculation variants, it is assumed that the mirror inaccuracies of the heliostats and concentrator are similar ($\sigma_c = \sigma_g = \sigma$). As mentioned above, it is considered that these mirror inaccuracies obey a normal law of errors. Under such an assumption, the dependence of the focal irradiance of the LSF on the mirror inaccuracy will be a function of one variable, which can be represented as the following empirical formula:

$$E_F = \frac{43578 E_0 R_{sc} R_{sg}}{1 - 0.034\sigma + 0.028\sigma^2} \left[\frac{W}{cm^2}\right]$$

where σ is the root-mean-square mirror inaccuracy per minute. The relative error of this formula is no more than 1.2%. A diagram of this dependence is shown in **Figure 5b**.

Let us consider the size of the focal spot in different cases. Usually, the value where the irradiance is 10% of the focal irradiance is selected as the size of the focal spot in solar concentrators. In the case of the LSF, for descriptive reasons and convenience of analysis, various levels of irradiance within 3–11 W/cm^2 were used as the contour (isometric) line of the focal spot. The contour lines of the focal spot for various mirror inaccuracies (E_0 = 700 W/m^2, $R_{sc} = R_{sg}$ = 0.6) are shown in **Figure 6**.

A certain idea of the character of the irradiance distribution for various mirror inaccuracies can be obtained from this, taking into account the value E_F from **Figure 5b**. It can be seen that the shape of the focal spot is not round, but a bit prolate along the horizontal axis. This is evidently related to the shape of the midsection of the concentrator. It can be also seen from **Figure 5b** that beginning from the second or third minute of the mirror inaccuracy, the character of the energy distribution (extension of the spot contour line) varies gradually and relatively uniformly as a function of σ.

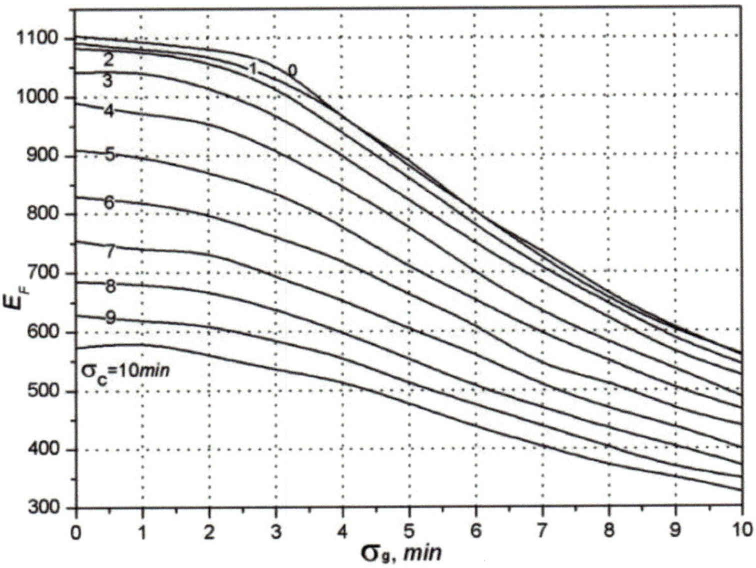

a

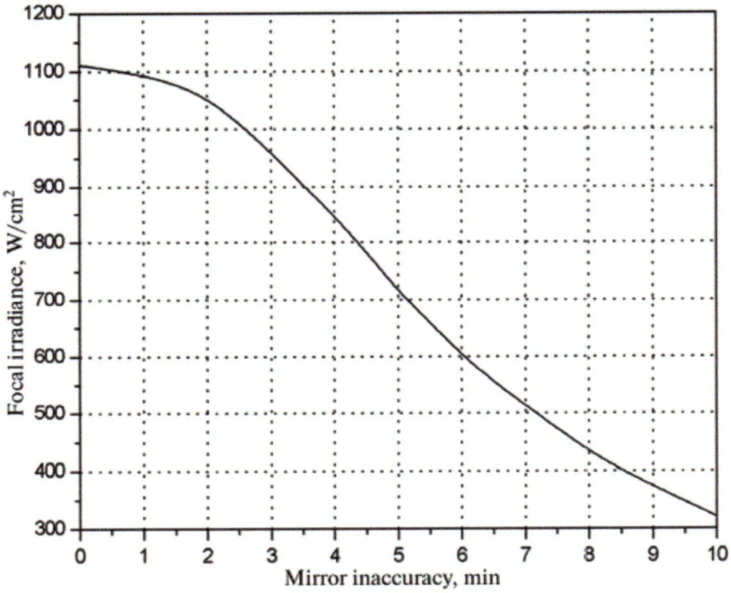

b

Figure 5.
(a) Dependence of focal irradiance E_F on the mirror inaccuracy of the heliostats σ_g and concentrator σ_c and (b) dependence of focal irradiance E_F on the mirror inaccuracy (case $\sigma_g = \sigma_c$).

Thus, as follows from the given analysis and without violating the generality of the results, the characteristic parameters of the LSP can be calculated, and its energy characteristics can be analyzed on their basis. On the basis of this data, the energy characteristics of the furnace can also be estimated for its other parameters. The following values were used as the characteristic parameters of the LSF: E_0 = 700 W/m^2 is the beam solar radiation, and $R_{sc} = R_{sg} = 0.6$ is the reflection coefficient of the mirrors of the concentrator and heliostats; the mirror inaccuracies of the concentrator and heliostats are $\sigma = 7$ ($\sigma = \sigma_c = \sigma_g$) angular minutes.

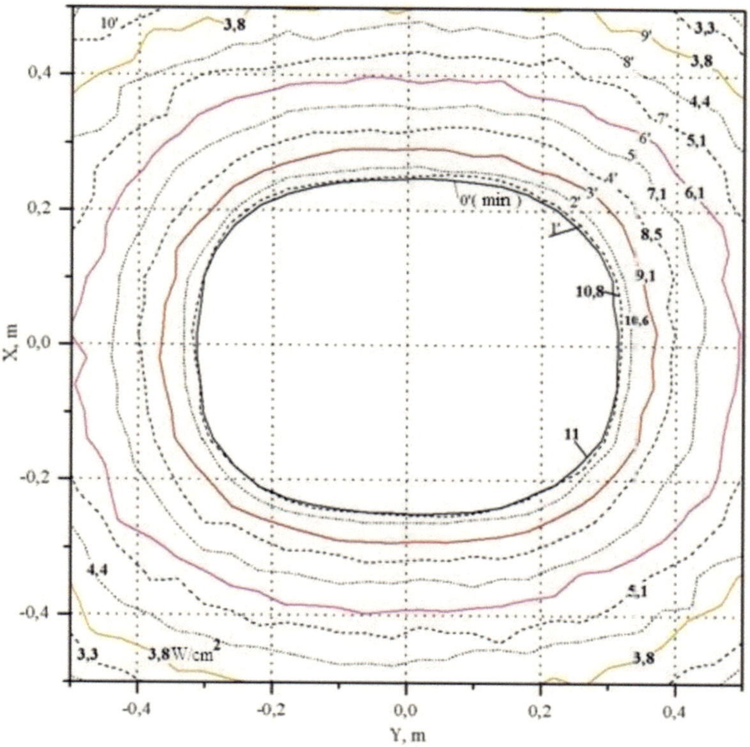

Figure 6.
Contour lines of focal spot.

Figure 7 shows the dependence (in the section) of the irradiance distribution in the focal zone of the LSF for the specified parameters with various inaccuracies of the reflecting mirrors.

Figure 8 shows the energy density distribution in the focal zone of the LSF in the form of isometric lines with equal irradiances and its three-dimensional

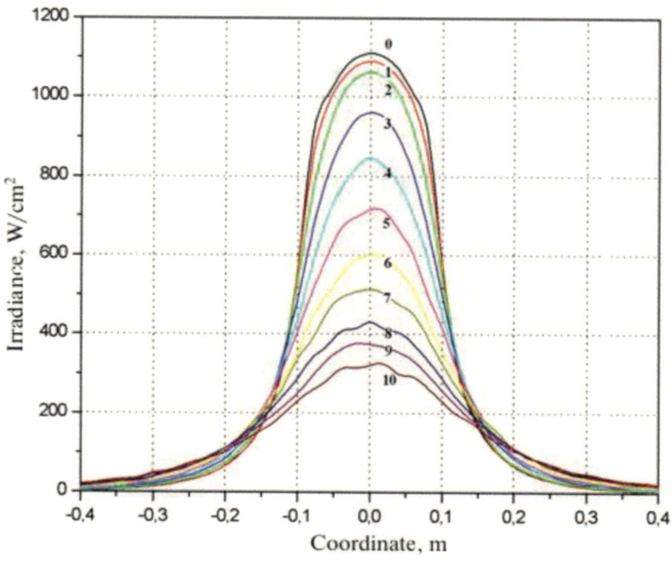

Figure 7.
Irradiance distribution in the section of focal zone of LSF.

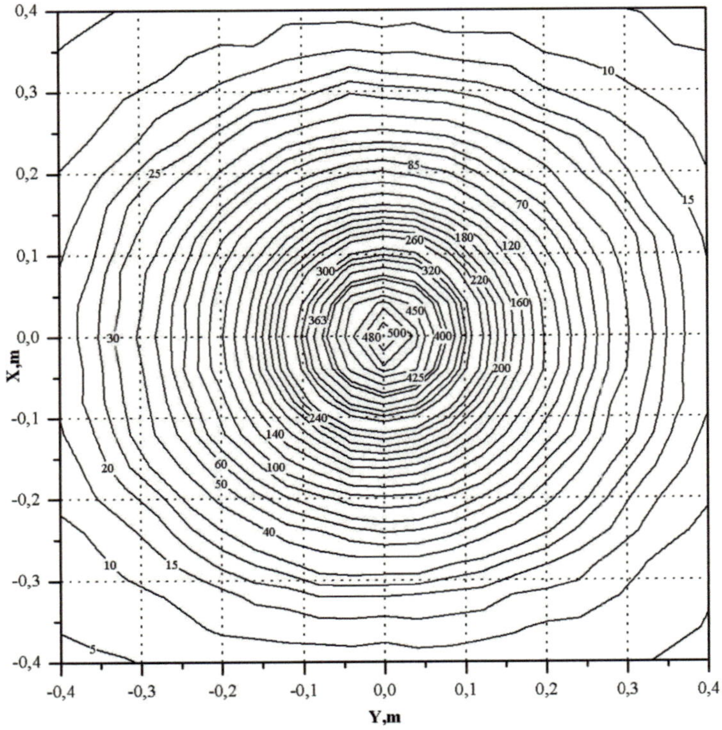

a

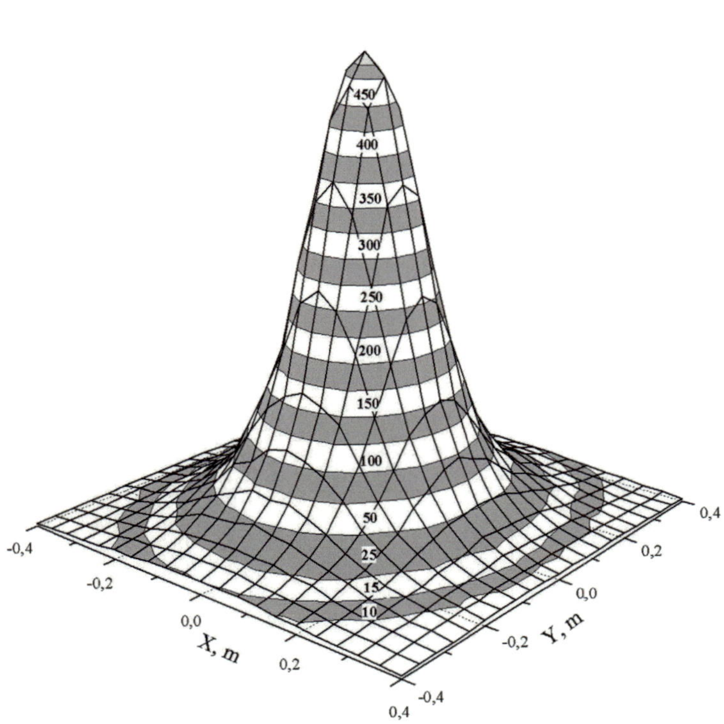

b

Figure 8.
(a) Energy density distribution from total system (isometric lines) and (b) energy density distribution from total system (three-dimensional).

representation specifying the irradiance levels. The maximum energy density at the center of the spot is 506 W/cm^2. The asymmetry of the focal spot can be also seen in the quantitative measures in **Figure 8**.

It should be noted that the furnace power in the selected parameters has a value of W_0 = 0.6•0.6•700 W/m^2•1906m^2 = 480.3кW. To check the correctness of the calculations, the amount of light energy per unit time (power) arriving at the focal zone of the LSF is determined in all calculations, and it was compared to the W_o value. It turned out that the difference between these values does not exceed 2–3%.

5.2.2 Representation of the calculated data by empirical formulas

Calculations are carried out to select the empirical formulas for the best approximation of the two-dimensional calculated data. The selection of the different types of functions shows that for this purpose, a two-dimensional Gaussian function (with six parameters) gives the best fit, which is given by

$$E(x,y) = E_0 + A \exp\left\{-\frac{1}{2}\left(\frac{x-x_c}{w_x}\right)^2 - \frac{1}{2}\left(\frac{y-y_c}{w_y}\right)^2\right\} \quad (7)$$

The LSF conditions with various inaccuracies of the reflecting mirrors are considered, and the parameters of the empirical formulas and standard root-mean-square deviations of irradiance are determined for each case according to the following formula:

$$E_\sigma = \sqrt{\frac{\sum_{i=1}^{N}\left(E_i - E(x_i,y_i)\right)^2}{N}}$$

where i is number of the current points, E_i is the calculated data, and N is point number, in our case, 2601. **Table 1** gives the determined parameters of the two-dimensional Gaussian functions and E_σ values for the various mirror inaccuracies of the heliostats and concentrator.

It is clear from **Table 1** that with an increase in mirror inaccuracies, the empirical formulas describe the calculated data better. Despite this, when using these formulas, it is necessary to make certain in each case that it is possible to use them. It should be noted that in all cases, the total power of the focal spot determined using the empirical formulas almost coincides with the power value determined by the calculated data. It also should be mentioned that in all cases, the sum of

σ, мин	E_0	A	x_c	w_x	y_c	w_y	E_σ
$\sigma_c = \sigma_g = 2$	4.71	1164.28	0.00006	0.074	0.00014	0.079	12.25
$\sigma_c = \sigma_g = 4$	7.02	840.87	0.000095	0.084	−0.00022	0.09	8.21
$\sigma_c = \sigma_g = 7$	9.24	474.82	−0.00031	0.106	0.00015	0.115	7.34
$\sigma_c = \sigma_g = 10$	10.53	290.26	0.0004	0.129	0.00058	0.139	6.10
$\sigma_c = \sigma_g = 12$	11.03	217.65	0.00056	0.143	0.00031	0.155	5.35
$\sigma_c = 7, \sigma_g = 5$	8.79	572.7	0.00054	0.098	0.0004	0.106	7.97
$\sigma_c = 7, \sigma_g = 4$	8.58	624.0	0.00039	0.094	0.000024	0.102	8.21
$\sigma_c = 5, \sigma_g = 3$	7.24	841.64	0.00031	0.084	−0.00012	0.09	8.42

Table 1.
Parameters of empirical formulas.

$(E_i - E(x_i, y_i))$ over all points is approximately zero. The empirical formulas were determined using the OriginPro8 graphics package (www.originLab.com).

5.2.3 Analysis of the power characteristics of the focal spot and average concentration

The value of the spot power and average radiation concentration in certain zones of the focal spot is important for some problems. The spot powers W in different square zones with the center at the focus of the concentrator (d is zone diameter) and average radiation concentrations E_{aver} in these zones are determined (**Figure 9**).

5.2.4 Analysis of the energy characteristics of certain LSF shelves

The energy characteristics of certain LSF shelves are also studied.

The three-dimensional irradiance distributions of certain shelves of heliostats are given for comparative analysis in **Figure 10**. The irradiance levels are specified in the belt lines. It is clear from **Figure 10** that the contribution of shelf 1 is the smallest from the viewpoint of the maximum irradiance, and the spot is more prolate along the X axis (vertically). The irradiance distributions from shelves 2, 3, and 8 have an approximately round symmetry. It is also seen that the focal spot of shelves 4, 5, 6, and 7 is more prolate along the Y axis (horizontally). The focal irradiance is maximum for shelf 4. The vertical section of the irradiance distribution from certain shelves is shown in **Figure 11**. The maximum values of irradiance by the heliostat shelves are as follows: 19, 41, 69, 93, 80, 80, 72, and 52 W/cm^2, and their sum is 506 W/cm^2. The quantitative characteristics of the irradiance distribution of certain shelves can be determined from the diagram.

Note that inclusion or exclusion of some individual shelves of heliostats in the tracking mode allows a certain set of possible irradiance distributions to be obtained. It is easy to see that the number of variants of inclusion/exclusion of the heliostat shelves N_p is 256. Actually,

$$N_p = \sum_{k=0}^{8} C_8^k = \sum_{k=0}^{8} \frac{8!}{(8-k)!k!} = (1 + 8 + 28 + 56 + 70 + 56 + 28 + 8 + 1) = 256$$

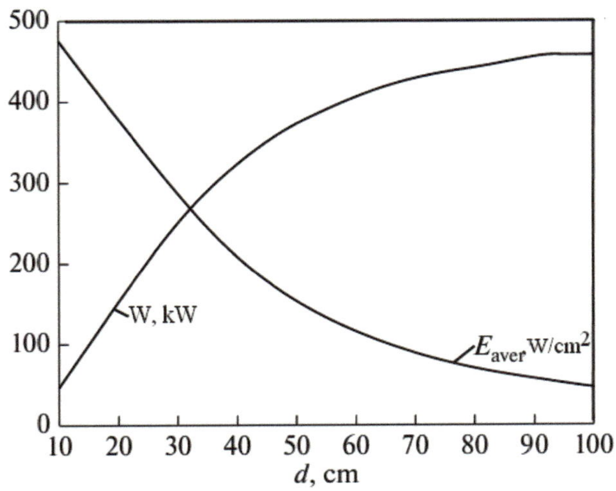

Figure 9.
Power and average concentration in different zones of focal spot.

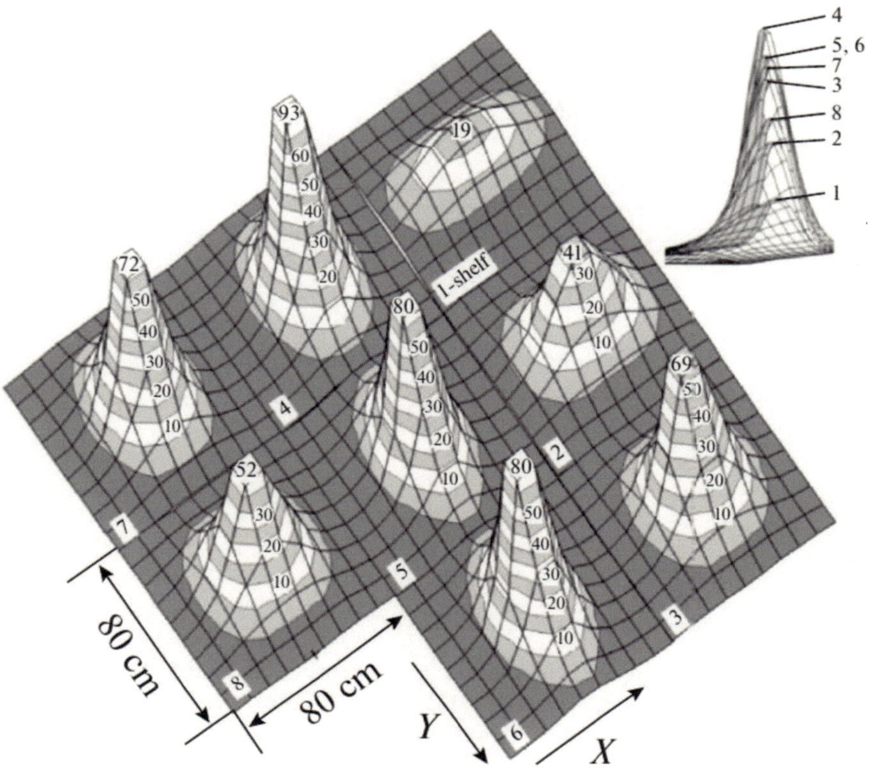

Figure 10.
Three-dimensional energy density distribution from certain shelves of LSF heliostats.

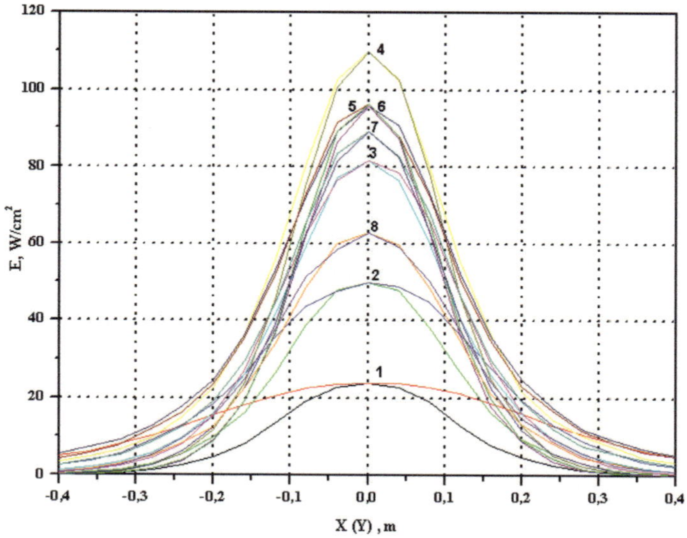

Figure 11.
Energy density distribution from certain shelves of LSF heliostats.

Figure 12 shows the dependence of the maximum irradiance (points in the diagram) of the distribution on the number *N* of the variant of inclusion/exclusion of the heliostat shelves. For example, the entry 01111111 means that all shelves except for first are engaged. Transfer from the number of the variant *N* to such a type of the entry is as follows: *N* is converted to the binary representation, and the

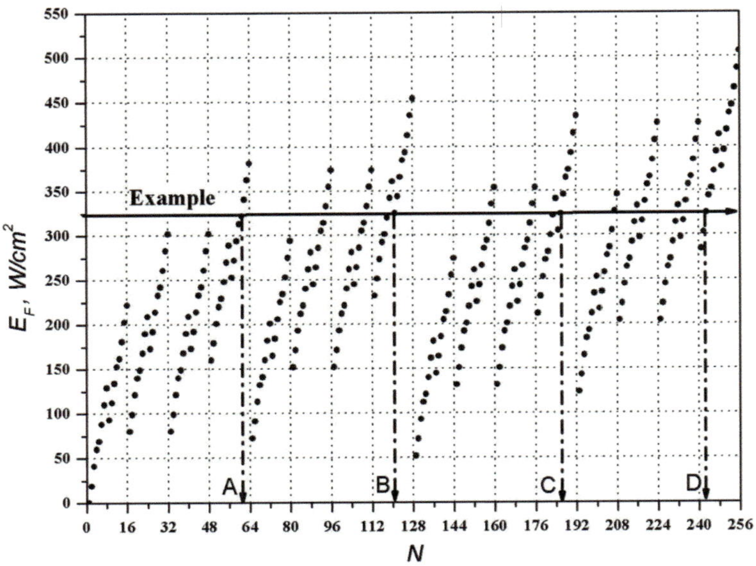

Figure 12.
Variants of inclusion/exclusion of shelves of LSF heliostats.

obtained number is read in reverse order. For example, $254_{10} = 11111110_2$, and the entry (combination) has the form 01111111. In the figure, the line *Example* corresponds to the required maximum density 325 W/cm^2, and such energy density is achieved using four variants: A, B, C, and D.

5.2.5 Heliostat rearrangement and its energy characteristics

Let us consider the energy characteristics of certain heliostats of the LSE. The focal irradiances and character of irradiance distribution of the heliostats significantly differ from each other. Analysis of the calculation results showed that the relative focal irradiance (normalized to the focal irradiance of the LSF) of the heliostats varies insignificantly with various mirror inaccuracies. In **Table 2**, this data is presented for 33 heliostats, taking symmetry into account.

Note that the maximum contribution of the heliostat to the focal irradiance is higher when the center of the heliostat is closer to the optical axis of the concentrator. This is why the heliostats can be grouped according to this principle. It can be expected that the shape of the focal spot of certain groups will be more symmetric.

No.	%	No.	%	No.	%	No.	%	No.	%
1	0.67	2	0.56	3	1.05	6	0.45	7	1.38
8	1.12	9	2.06	13	0.44	14	0.79	15	2.18
16	1.74	17	3.21	22	0.79	23	1.04	24	2.96
25	2.41	26	4.34	31	0.93	32	1.2	33	3.45
34	2.25	39	1.02	40	0.97	41	3.39	42	2.17
47	0.8	48	1.04	49	2.96	50	2.34	55	0.25
56	1.31	57	1.24	58	2.55				

Table 2.
The relative focal irradiance of the heliostats.

Based on the analysis of illumination of the concentrator by the heliostats, four groups of heliostats are distinguished. The heliostats with a close distance from the optical axis of the concentrator are combined in each group. The numbers of heliostats in these groups are as follows:

Group 1: 26, 34, 35, 42, 43, and 5 heliostats (1–5, rectangle).

Group 2: 16, 17, 18, 24, 25, 27, 28, 33, 36, 41, 44, 49, 50, 51, 52, 58, 59, and 17 heliostats (6–22, circles).

Group 3: 8, 9, 10, 15, 19, 23, 29, 32, 37, 40, 45, 48, 53, 57, 60, and 15 heliostats (23–37, triangles).

Group 4: 1, 2, 3, 4, 5, 6, 7, 11, 12, 13, 14, 20, 21, 22, 30, 31, 38, 39, 46, 47, 54, 55, 56, 61, 62, and 25 heliostats (38–62, rhombs).

The new number of the heliostats is given in brackets. In **Figure 3**, heliostats belonging to one group are designated, respectively, by rectangles, triangles, circles, and rhombs. As the analysis of the numerical calculations has shown, the shape of the focal spot of certain groups of heliostats is almost symmetric, except for group 4, where a slight asymmetry is observed. This is why the energy density distributions from certain groups in the horizontal section (Y axis) are shown in **Figure 13**.

In the general case, to determine the necessary possible focal irradiance (or the required irradiance), it suffices to construct a diagram of the dependence of the cumulative sum of the focal irradiance on the number of the heliostat, i.e.,

$$y(n) = \sum_{i=1}^{i=n} E_F(i), \quad n = 1 - 62$$

This diagram is similar to that in **Figure 12** in the character of construction; only the new numbers of heliostats are measured on the horizontal axis. A diagram of function $y(n)$ for such a new group of heliostats is shown in **Figure 14**.

5.2.6 Analysis of the focal spot of the individual heliostats

The authors performed a comparative analysis of the energy density distributions from certain heliostats. The focal irradiances of certain heliostats can be determined from **Table 2**, knowing the total focal irradiance of the LSF, 506 W/m^2. The isometric line corresponding to a value of 1 W/cm^2 is selected from the

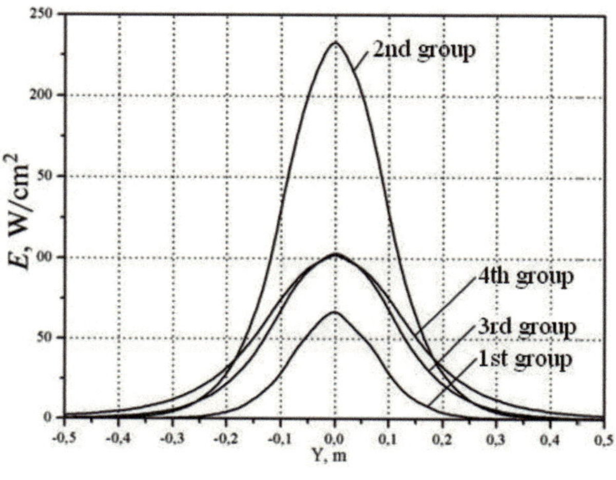

Figure 13.
Energy density distribution from certain groups of heliostats.

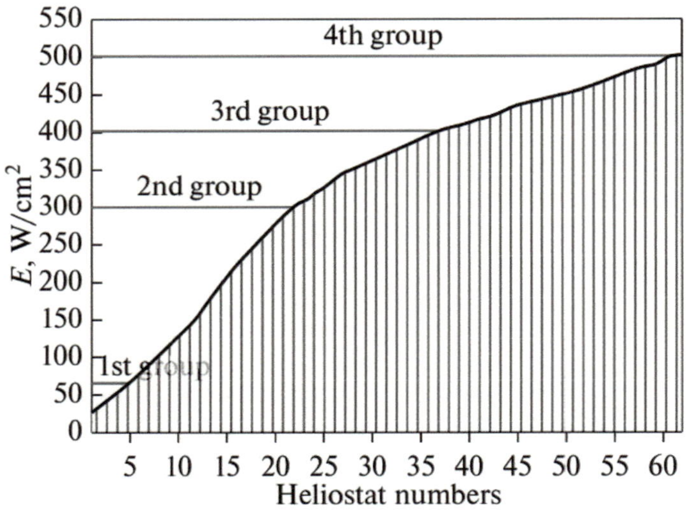

Figure 14.
Increasing energy density from certain groups of heliostats.

two-dimensional diagram of the irradiance distribution from a certain heliostat represented as isometric lines of a similar irradiance. It is clear that this isometric line corresponds to the contour line of the focal spot of the considered heliostat. The contour lines of the spots of certain heliostats (33 heliostats) are shown in **Figure 15a** and **b**. For descriptive reasons, so as not to overload the diagram, the heliostats are conditionally divided into two groups, central (left) and extreme (right). The contour line of the focal spot of the selected heliostat can be accurately traced from the diagram.

The character of the energy density distribution can be estimated if the maximum irradiance and geometry of the focal spot of the heliostat are known.

5.2.7 Analysis of the changes of the energy distribution within a day

One of the most important problems is the change in the focal energy density distribution within a day, occurring due to the daily variation in beam solar radiation. As an example, **Figure 16** shows a diagram characterizing the change in the focal energy density distribution along the horizontal axis of the LSF. For the beam

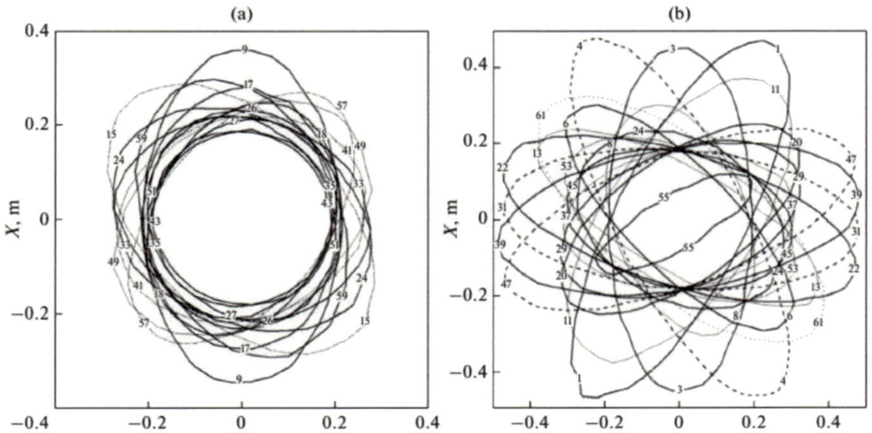

Figure 15.
Contour lines and orientation of focal spots of certain heliostats of LSF (Figures in contours - numbers of heliostats)

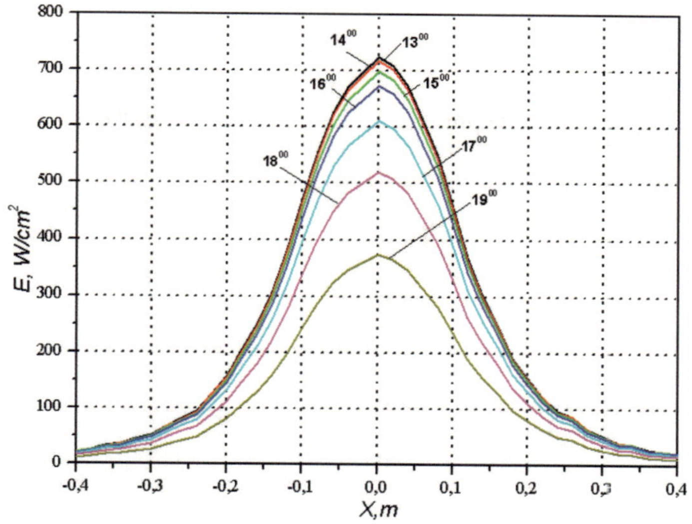

Figure 16.
Variation in energy density within 1 day.

solar radiation, a clear sunny day is selected according to data from the Parkent meteorological station: 18 June 2013. It can be seen from **Figure 16** that within the time interval of 10:00–16:00, i.e., within 6 h (13:00 is approximately midday), the character of the energy density distribution does not change significantly. It should be noted that a similar analysis should be carried out for other periods of time.

6. Conclusion

Let us note, lastly, that the performed theoretical and design analyses of the LSF energy characteristics and their revealed features, general patterns, and presented detailed data allow to correctly determine the configuration parameters of the LSF for various required process modes in the focal zones of the furnace. The long-term experience in application of the LSF for synthesis and thermal processing of various high-temperature materials has shown that the accuracy in implementing certain process modes significantly affects the final characteristics of the product. This is why the obtained results have important practical value for efficient operation of the LSF of the Academy of Sciences of the Republic of Uzbekistan.

Author details

Akbarov Rasul
Institute of Materials Science of the Academy of Sciences of the Republic of Uzbekistan, Tashkent, Uzbekistan

*Address all correspondence to: aryu12@mail.ru

References

[1] Rakhimov RK. Synthesis of functional ceramics on the BSP and development on its base. Computational Nanotechnology. 2015;**3**:11-25

[2] Flamant G, Ferriere A, Laplaze D, Monty C. Solar processing of materials: Opportunities and new frontiers. Solar Energy. 1999;**66**(2):117-132

[3] Trombe F, Gion L, Royere C, Robert JF. First results obtained with the 1000-kW solar furnace. Solar Energy. 1973;**15**: 63-66

[4] Alvarez L, Guillard T, Olalde G, Rivoire B, Robert JF, Bernier P, et al. Large scale solar production of fullerenes and carbon nanotubes. Synthetic Metals. 1999;**103**(1–3): 2476-2477

[5] Flamant G, Balat-Pichelin M. Elaboration and testing of materials using concentrated solar energy. In: Blanco Galvez J, Rodriguez SM, Delyannis E, Belessiotis VG, Bhattacharya SC, Kumar S, editors. Solar Energy Conversion and Photoenergy Systems. Vol. I. United Kingdom: Eolss Publishers Co., Ltd./ UNESCO; 2010. pp. 363-389

[6] Riskiev TT, Abdurakhmanov AA, Khodjaev RA, Akbarov RY. The outlook for technical hydrogen production in the big solar furnace. Applied Solar Energy. 2003;**38**(4):49-54

[7] Fernandez-Gonzalez D et al. Concentrated solar energy applications in materials science and metallurgy. Solar Energy. 2018;**170**:520-540

[8] Herranz G, Rodriguez GP. Uses of concentrated solar energy in materials science. In: Rugescu RD, editor. Solar Energy. Croatia: InTech; 2010. pp. 146-170

[9] Azimov SA. Research-industrial complex Sun. Bimirror polyheliostat solar furnace with a thermal power of 1000 kW. Geliotekhnika. 1987;**6**:3-6 in Russian

[10] Abdurakhmanov AA, Akbarov RY, Gulamov KG, Riskiev TT. Operating experience of a big solar furnace 1000 kW in power. Applied Solar Energy. 1993;**34**(1):29-33

[11] Trombe F, Vinh ALP. Thousand kW solar furnace, built by the National Center of Scientific Research, in Odeillo (France). Solar Energy. 1973; **15**(1):57-61

[12] Riskiev TT, Suleimanov S.Kh. Double mirror polyheliostat solar furnace of 1000 kW thermal power. Solar Energy Materials. 1991;**24**(1–4): 625-632

[13] Akbarov RY, Paizullakhanov MS. Characteristic features of the energy modes of a large solar furnace with a capacity of 1000 kW. Applied Solar Energy. 2018;**54**(2):99-109

[14] Abdurakhmanov AA, Akbarov RY, Sobirov YB, Yuldashev AA. Method of measurement and control of optical and geometric characteristics of mirrors and glasses. Applied Solar Energy. 2003; **39**(1):62-65

[15] Abdurakhmanov AA, Akbarov RY, Sobirov YB. Application of the system of technical vision in the big solar furnace. Applied Solar Energy. 1998; **34**(1):38-40

[16] Abdurakhmanov AA, Akbarov RY, Sobirov YB, Saribaev AS. A photometer for measuring the density of concentrated solar radiation. Applied Solar Energy. 2000;**36**(2):72-75

[17] Abdurakhmanov AA, Akbarov RY, Sobirov YB. Analysis of operating characteristics of various smelting

furnaces on a large solar furnace. Applied Solar Energy. 2008;**44**(1):24-27

[18] Abdurakhmanov AA, Akbarov RY, Sarybaev AS, Yuldashev AA. The automated control system of the heliostat field of the big solar furnace. Applied Solar Energy. 1998; **34**(1):34-37

[19] Atabaev IG, Akhatov ZS, Mukhamediev ED, Zievaddinov Z. Modernization of an automated controlling system for heliostat field of big solar furnace. Applied Solar Energy. 2016;**52**(3):220-225

[20] Faiziev SA, Sobirov YB. Measurements of solar resources in Uzbekistan. Applied Solar Energy. 2017; **53**(1):57-60

[21] Abdurakhmanov AA, Sobirov Yu B, Paizullakhonov MS, Orlov SA. Results of actinometric measurements at location of LSF with thermal capacity of 1000 kW. Applied Solar Energy. 2012; **48**(3):228-231

[22] Guillot E, Rodriguez R, Boullet N, Sans J-L. Some details about the third rejuvenation of the 1000 kWh solar furnace in Odeillo: Extreme performance heliostats. In: SolarPACES; 2017; Santiago, Chile: AIP Conference Proceedings; 2018. DOI: 10.1063/ 1.5067052

[23] Zakhidov RA, Weiner AA, Umarov G. Theory and Calculation of Applied Solar Energy Concentrating Systems (English Ed). Vadodara, India: Gujarat Energy Development Agency; 1992. p. 146

[24] Grilikhes VA, Matveev VM, Poluetkov VP. Solar high-temperature heat sources for spacecraft. Moscow: Mechanical Engineering; 1975. 248 p. (in Russian)

[25] Abdurakhmanov AA, Akbarov RY, Riskiev TT, Lewandowski A. On calculation of optical-energetic characteristics of double mirror concentrating systems. Applied Solar Energy. 2002;**38**(2):71-77

[26] Klychev Sh I, Bakhramov SA, Zakhidov RA, Akbarov RY, Klycheva MS. Solar energy concentrators—Errors of numerical calculation of the irradiance integral for solar paraboloid concentrators. Applied Solar Energy. 2005;**41**(2):55-58

[27] Abdurakhmanov AA, Akbarov RY, Sarybaev AS, Yuldashev AA. A tower-type concentrator based on heliostats of a big solar furnace 1000 kW in power. Applied Solar Energy. 1999;**35**(2):60-65

[28] Akbarov RY, Pirmatov II, Riskiev TT. Specific features of the focal spot formation in big solar furnaces with polyheliostat systems. Applied Solar Energy. 1998;**34**(6):29-33

[29] Akbarov RY, Pirmatov II, Riskiev TT. Monitoring the energy density distribution in the focal zone of the big solar furnace. Applied Solar Energy. 1998;**34**(1):47-51

[30] Korn GA, Korn TM. Mathematical Handbook for Scientists and Engineers: Definitions, Theorems, and Formulas for Reference and Review. Mineola, New York: Dover publications, Inc.; 2000. 1151 pp

[31] Levin BR. Theoretical Basics of Statistical Radiotechnics, Book 1. Moscow: Soviet Radio Publ. Co.; 1969. 752 pp. (in Russian)